Digitalizing Material Innovations for Net-Zero Applications

Digitalizing Material Innovations for Net-Zero Applications navigates the critical intersection of advanced materials science, cutting-edge digital engineering, and global sustainability goals. It explores how digital tools are fundamentally transforming the way we innovate, develop, and deploy materials essential for achieving net-zero emissions across diverse sectors, from energy systems to smart cities and industrial processes.

Drawing on expert contributions, this book provides in-depth insights into key areas such as artificial intelligence–driven design of CO_2 capture materials and membranes, strategies for decarbonizing the built environment, and the transformative potential of Power-to-X technologies. It addresses the growth potential and commercialization barriers for next-generation materials, offering practical solutions and real-world case studies, including the application of digital systems for sustainable manufacturing in MSMEs. This unique blend of material science and digital solutions offers a holistic perspective on accelerating decarbonization.

This resource is designed for industry professionals and experts working on the decarbonization agenda in the fields of materials processing, manufacturing, and industrial sustainability initiatives.

Faiz Iqbal is a Senior Lecturer in Mechanical Engineering at the University of Lincoln, UK. His research focuses on smart manufacturing, Industry 4.0, and cyber-physical systems, with a particular emphasis on digital transformation for micro- and small-scale enterprises (MSMEs). He leads and co-leads several interdisciplinary research projects, including British Council and Innovate UK-funded initiatives on net-zero manufacturing, AI-based quality inspection, and sustainable automation. Dr. Iqbal holds a PhD in Manufacturing Automation from the Indian Institute of Technology Delhi. He has developed advanced automation frameworks and novel CNC-integrated finishing technologies, and his work on cyber-physical disinfectant manufacturing during the COVID-19 pandemic was nationally recognized in the UK with the COVID-19 Engineering Medal and a Scottish Knowledge Exchange Award nomination.

Suneela Sardar is a Senior Lecturer in Chemical Engineering at the University of Lincoln, UK, and a passionate Chemical Engineer dedicated to advancing sustainable energy technologies and mitigating environmental challenges. She completed her PhD in Chemical Engineering at the University of Leeds, UK. Dr. Sardar has also worked on many consultancy projects on flow behaviour of polymeric materials and surface modification of particles. As a Postdoctoral Research Fellow at the University of Leeds, she worked on sustainable minerals processing for the extraction of valuable materials to support the foundation industries. At Lancaster University, she worked in the radioactive lab to research the molten salts for nuclear energy applications and simulated their behaviour.

Digitalizing Material Innovations for Net-Zero Applications

Edited by

Faiz Iqbal and Suneela Sardar

CRC Press
Taylor & Francis Group
Boca Raton London New York

CRC Press is an imprint of the
Taylor & Francis Group, an **informa** business

Designed cover image: Faiz Iqbal

First edition published 2026
by CRC Press
2385 NW Executive Center Drive, Suite 320, Boca Raton FL 33431

and by CRC Press
4 Park Square, Milton Park, Abingdon, Oxon, OX14 4RN

CRC Press is an imprint of Taylor & Francis Group, LLC

ISBN: 978-1-032-94903-1 (hbk)
ISBN: 978-1-032-94904-8 (pbk)
ISBN: 978-1-003-58230-4 (ebk)

DOI: 10.1201/9781003582304

Typeset in Times
by codeMantra

Contents

Editors

Dr. Faiz Iqbal is a Senior Lecturer in Mechanical Engineering at the University of Lincoln, UK. His research focuses on smart manufacturing, Industry 4.0, and cyber-physical systems, with a particular emphasis on digital transformation for micro- and small-scale enterprises (MSMEs). He leads and co-leads several interdisciplinary research projects, including British Council and Innovate UK-funded initiatives on net-zero manufacturing, AI-based quality inspection, and sustainable automation. Dr. Iqbal holds a PhD in Manufacturing Automation from the Indian Institute of Technology Delhi and has previously held academic and research appointments at Mahindra University (India) and the University of Edinburgh (UK). He has developed advanced automation frameworks and novel CNC-integrated finishing technologies, and his work on cyber-physical disinfectant manufacturing during the COVID-19 pandemic was nationally recognized in the UK with the COVID-19 Engineering Medal and a Scottish Knowledge Exchange Award nomination. With a strong interest in translational research and industry collaboration, Dr. Iqbal has worked with a wide range of industrial partners and has supervised award-winning student projects in the areas of digifacturing, intelligent automation, and sustainable systems design. He regularly publishes in leading journals and serves as a reviewer for several high-impact international publications in the fields of manufacturing, robotics, and applied engineering.

Dr. Suneela Sardar is a passionate Chemical Engineer dedicated to advancing sustainable energy technologies and mitigating environmental challenges. She did her PhD in Chemical Engineering at the University of Leeds and pursued on sustainable methods and materials for solid oxide fuel cells. During her PhD, Dr. Sardar also worked on many consultancy projects on flow behaviour of polymeric materials and surface modification of particles. As a Postdoctoral Research Fellow at the University of Leeds, she worked on sustainable minerals processing for the extraction of valuable materials to support the foundation industries. At Lancaster University, she worked in the radioactive lab to research the molten salts for nuclear energy applications and simulated their behaviour. Currently, Dr. Sardar is a Senior Lecturer in Chemical Engineering at the University of Lincoln and involved in teaching and research activities. She is closely following and involved in securing the materials for Fusion Energy development in Lincolnshire, a UK initiative for clean and sustainable energy production to achieve the net-zero targets. Having a strong interest in research, she is working on interdisciplinary research topics to tackle pressing global challenges.

Contributors

Nor Naimah Rosyadah Ahmad
Faculty of Applied Sciences
Universiti Teknologi MARA
Selangor, Malaysia

Shadab Ahmad
School of Mechanical and Electrical Engineering
Changchun University of Technology
Changchun, China

Salem K. Brini Ahmed
Sebha University
Sabhā, Libya

Usman Ali
Department of Chemical Engineering
University of Engineering and Technology
Lahore, Pakistan

Aliyu M. Aliyu
School of Engineering and Physical Sciences
University of Lincoln
Lincoln, United Kingdom

Benjamin Anim
Department of Mechanical Engineering
Kumasi Technical University
Kumasi, Ghana

Edward Antwi
Institute of Renewable Energy Systems
University of Applied Science
Berlin, Germany

Yahaya D. Baba
School of Water, Energy and Environment (SWEE)
Cranfield University
Cranfield, United Kingdom

Abhishek Bhatnagar
Indian Institute of Technology
New Delhi, India

Divya Bhatnagar
Indian Institute of Technology
Hyderabad, India

Amr Elkady
Titan Cement Egypt
Alexandria and Bani Suef Cement Companies
Alexandria, Egypt

Faiz Iqbal
School of Engineering and Physical Sciences
University of Lincoln
Lincoln, United Kingdom

Khuram Maqsood
Department of Chemical Engineering
University of Jeddah
Jeddah, Saudi Arabia

Noha A. Mostafa
Department of Industrial Engineering
Zagazig University
Zagazig, Egypt
Mechanical Engineering Department
The British University in Egypt
Cairo, Egypt

Rizwan Nasir
Department of Chemical Engineering
University of Jeddah
Jeddah, Saudi Arabia

Flavio Odoi-Yorke
Department of Renewable Energy Technology
Cape Coast Technical University
Cape Coast, Ghana

Joseph X.F. Ribeiro
Department of Oil and Gas Engineering
Kumasi Technical University
Kumasi, Ghana

Eleonora Ricci
Institute for Materials and Processes
University of Edinburgh
Edinburgh, United Kingdom

Suneela Sardar
School of Engineering and Physical Sciences
University of Lincoln
Lincoln, United Kingdom

Muhammad Shahbaz
Net Zero Industry Innovation Centre
Teesside University in Egypt
Teesside, Egypt

Ziqi Shen
Energy and Sustainability Theme
Cranfield University
Bedford, United Kingdom

Alhassan Haruna Umar
School of Engineering and Physical Sciences
University of Lincoln
Lincoln, United Kingdom

Siqi Wang
Energy and Sustainability Theme
Cranfield University
Bedford, United Kingdom

Yongliang Yan
School of Engineering
Newcastle University
Newcastle upon Tyne, United Kingdom

Siti Zulaikha Mohd Yusof
Faculty of Applied Sciences
Universiti Teknologi MARA
Selangor, Malaysia

Preface

In an era defined by unprecedented environmental challenges, the global imperative to achieve net-zero emissions has become the defining mission of our time. This ambitious goal necessitates a fundamental transformation across all sectors, driven by relentless innovation in both materials science and digital technologies. It is at this crucial intersection that this book, *Digitalizing Material Innovations for Net-Zero Applications,* finds its purpose.

The genesis of this volume stems from a recognition of the accelerating convergence between advanced materials development and the pervasive power of digitalization. Traditional approaches to material discovery and optimization, while foundational, often struggle to keep pace with the urgent demands of climate change mitigation. Digital tools ranging from computational modelling and artificial intelligence to real-time monitoring and data analytics offer transformative capabilities, enabling faster discovery, more efficient processes, and smarter deployment of sustainable materials. We observed a growing need for a comprehensive resource that not only showcases the latest advancements in materials for decarbonization but also critically examines how digital solutions are enabling and accelerating their journey from concept to industrial application.

This book is meticulously structured into three distinct yet interconnected sections, designed to provide a holistic understanding of this dynamic field:

- **Section I: Materials for Decarbonization** lays the scientific groundwork, delving into specific material innovations critical for reducing carbon footprints. Chapters in this section explore the chemistry, performance, and application of cutting-edge materials for CO_2 capture, such as amine-based systems, and the computational advancements in membrane technologies for emissions reduction.
- **Section II: Growth Potential and Associated Barriers in Commercialization** shifts focus to the practical realities of implementing these innovations. This section critically analyses the market opportunities for sustainable materials; identifies the technical, economic, and policy barriers hindering their widespread adoption; and presents compelling case studies, including the potential of solar energy in emerging economies.
- **Section III: Digitalization in Materials for Net-Zero Applications** brings the core theme to the forefront, showcasing how digital tools are the indispensable enablers of this transition. Chapters here explore the transformative role of Power-to-X technologies in integrating renewable energy and propose novel digital systems for real-time emissions monitoring and sustainability awareness in manufacturing, particularly for small- and medium-sized enterprises.

As editors, we believe that the journey to net zero will be fundamentally shaped by our ability to leverage the synergy between material science and digital engineering.

This book serves as a testament to the ingenuity and dedication of researchers and practitioners worldwide who are at the forefront of this endeavour. It highlights not only the remarkable progress being made but also the persistent challenges that require continued interdisciplinary collaboration and innovative thinking. We are confident that by embracing digital transformation, we can unlock unprecedented opportunities to create a more sustainable and resilient future.

We extend our sincere gratitude to all the contributing authors for their invaluable expertise and tireless efforts in crafting their respective chapters. Their diverse perspectives and deep insights form the backbone of this publication. We also wish to thank the reviewers for their constructive feedback, which significantly enhanced the quality and coherence of this volume. Finally, our appreciation goes to the team at Taylor & Francis/CRC Press for their unwavering support and guidance throughout the publishing process.

We hope that this book will serve as a valuable resource for its readers, inspiring further research, fostering new collaborations, and ultimately contributing to the global mission of achieving net-zero emissions.

Editors:
Dr. Faiz Iqbal
Dr. Suneela Sardar
School of Engineering and Physical Sciences
University of Lincoln, Lincoln, United Kingdom – July 2025

1 Digital Frontiers of Materials Innovation for Net-Zero Applications

Suneela Sardar

Imagine a world where the materials that build our cities, power our vehicles and store our energy are not just discovered in labs, but are instead designed, simulated and optimised through integrated digital platforms. This is not a distant vision; it is the future of materials innovation and happening now. Due to net-zero urgency, the pressure to develop sustainable, high-performance materials has never been greater. But traditional methods of discovery are too slow, costly and uneven to meet the urgency of the climate crisis. That is why digitalisation steps in as a transformation.

This book explores the intersection of materials science, digital engineering and sustainability. It is organised for researchers, engineers, technologists and decision-makers who are actively shaping the future of materials for net-zero applications, whether in energy storage, lightweight structures, carbon capture or beyond. Through real-world examples, emerging frameworks and practical insights, we will examine how digital tools are transforming the way we innovate, collaborate and deliver impact.

The focus is on digital frontiers of materials innovation for net-zero applications by introducing the transformative theme of digitalising materials innovation, stressing its critical role in reshaping how materials are developed, optimised and utilised across industries. With the global race toward more sustainable practices, the integration of digital tools into materials science offers unique opportunities to accelerate discovery, reduce costs and performance enhancement. This transformation is particularly demanding in the context of achieving net-zero emissions, where materials innovation plays a critical role. From lightweight composites to sustainable alternatives and advanced semiconductors, innovative materials are central to climate solutions.

Net Zero refers to the balance between the amount of greenhouse gases emitted into the atmosphere and the amount removed from it. Achieving net zero means reducing emissions as much as possible and offsetting any remaining emissions through carbon removal strategies like reforestation or carbon capture technologies [1]. This concept is central to combating climate change, as it directly addresses the root cause: the accumulation of greenhouse gases in the atmosphere.

The urgency of reaching net zero is underscored by the accelerating impacts of climate change: rising sea levels, extreme weather events and biodiversity loss. According to the Intergovernmental Panel on Climate Change, global emissions

DOI: 10.1201/9781003582304-1

must be halved by 2030 and reach net zero by 2050 to limit global warming to 1.5°C above pre-industrial levels [2]. This timeline is critical to avoid irreversible damage to ecosystems and human societies.

Governments, corporations and institutions worldwide are adopting net-zero targets, recognising that the transition to a low-carbon economy is not only an environmental imperative but also an economic opportunity. The path to net zero involves systemic changes in energy, transportation, industry and land use, supported by innovation, policy and international cooperation. Figure 1.1 shows the timeline of achieving net-zero emissions.

Large corporations may possess the resources for digital transformation, but small- and medium-sized enterprises (SMEs) have difficulties such as limited resources to access infrastructure and shortage of skills [3]. Despite these challenges, SMEs are important for national economies and carbon reduction goals [4]. Therefore, this chapter also explores how SMEs can be supported to promote digitalisation for sustainable materials innovation and achieving climate objectives. Recent research shows that digitalisation and net-zero innovation in SMEs are not only compatible but mutually strengthening [5]. Low-cost digital tools for monitoring energy, quality and waste offer practical pathways for SMEs to engage in Industry 4.0 and climate action simultaneously [6].

Digitalisation in the context of materials innovation refers to the application of digital technologies, such as data analytics, artificial intelligence (AI), machine learning (ML) and automation to accelerate the discovery, design, testing and deployment of new materials [7]. The global imperative to achieve net-zero greenhouse gas emissions has made sustainability central to research and industrial strategies. Net zero refers to the balance between the greenhouse gases emitted and those removed from the atmosphere, and it has become a critical target for governments, industries and innovators worldwide [4]. Understanding the materials innovation lifecycle is essential for achieving these goals. This lifecycle typically includes stages such as discovery, development, scale-up, production, use and end-of-life management.

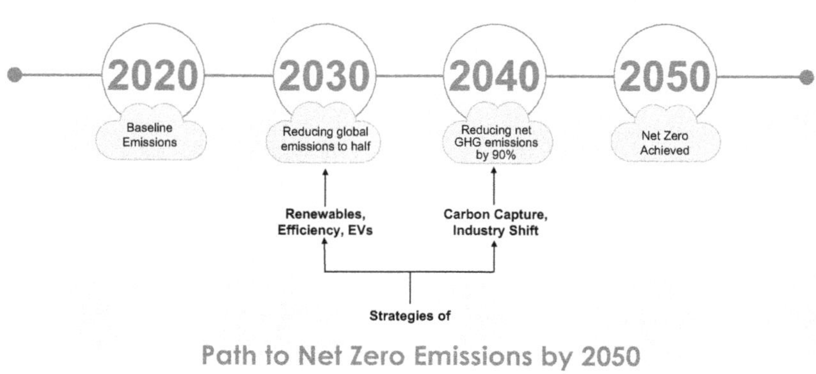

FIGURE 1.1 Net-zero timeline.

Digitalisation enhances each of these stages by enabling predictive modelling, virtual testing and improved lifecycle analysis [5]. The integration of AI and automation can significantly reduce the time and cost associated with experimental iterations and allow for more efficient resource use [6]. Moreover, digital tools provide SMEs with access to capabilities previously reserved for large enterprises, thus democratising innovation and sustainability. By embedding data and intelligence into the materials innovation process, digitalisation not only enhances performance but also aligns industrial output with long-term environmental and economic objectives [8].

The integration of advanced digital tools is reshaping the landscape of materials science by enabling smarter, faster and more cost-effective innovation processes. Simulation and modelling software, such as density functional theory and molecular dynamics, play a fundamental role in the virtual design of materials, allowing researchers to predict properties and behaviours before physical synthesis [7]. These tools drastically reduce the trial-and-error associated with traditional experimentation and facilitate rapid screening of materials for specific applications. Complementing simulation, AI and ML have emerged as powerful methods for materials discovery. By analysing large datasets from experiments and literature, ML algorithms can identify hidden patterns and propose novel material compositions with desired performance characteristics [6].

In the scope of manufacturing and materials processing, digital twins and the Internet of Things (IoT) provide real-time monitoring and control over production environments. Digital twins can simulate changes in materials or processes, optimise performance and reduce waste [8]. IoT sensors integrated into machinery provide continuous data streams, enabling predictive maintenance and quality assurance.

SMEs, which form the backbone of many economies, often struggle to engage in this transition due to limited resources and digital capabilities. Recognising this, governments and international institutions are increasingly offering policy and funding support to help SMEs adopt digital technologies. Initiatives include financial incentives, digital transformation roadmaps and innovation hubs that connect SMEs with expertise and infrastructure [9]. These efforts aim to democratise access to digital innovation, ensuring that SMEs contribute to, and benefit from, the green industrial transition.

Achieving net-zero emissions in SMEs sector revolved around some major parameters as shown in Figure 1.2. Manufacturing cycle in μ-SMEs involves supply chain, production stage and waste management. Assessing the energy consumption and emissions is very important as they happen at all the stages of manufacturing process. Carbon footprint analysis of manufacturing process takes into consideration those factors which impact environment, including raw materials transportation and processing, production process, distribution of products and end-of-life treatment of products [10,11].

For SMEs, which often lack the infrastructure of large corporations, cloud-based platforms and collaborative digital tools offer accessible and scalable solutions. These platforms enable remote data analysis, storage and team collaboration across geographical boundaries, lowering the entry barrier for digital transformation [12]. Furthermore, many cloud platforms now include built-in AI tools, simulation modules and interfaces for digital twin integration, making advanced materials innovation

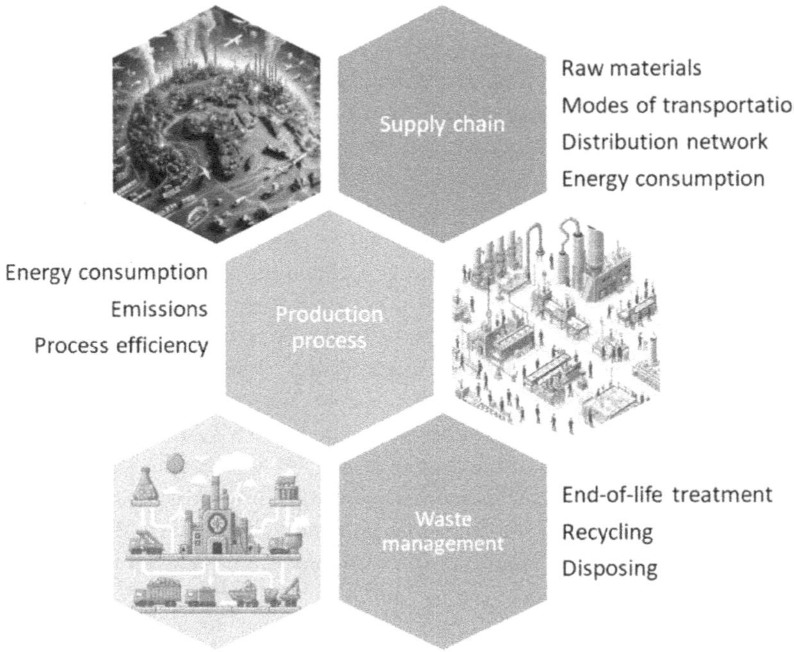

FIGURE 1.2 Carbon footprint analysis in SMEs.

more inclusive and efficient. Overall, digital tools empower SMEs to participate in cutting-edge research and sustainable manufacturing, bridging the innovation gap between small and large enterprises.

Importantly, digitalisation in materials innovation aligns closely with global climate goals such as the Paris Agreement and national net-zero strategies. It not only supports decarbonisation but also enhances industrial competitiveness and resilience. As countries push for greener economies, the strategic integration of digital tools into materials innovation becomes not just beneficial, but essential.

This book comprises a substantial number of chapters, each addressing key dimensions of materials innovation within the broader context of net-zero applications. *Chapter 2* explores how digital tools accelerate the design, screening and optimisation of amine-based materials for carbon capture technologies. *Chapter 3* examines how digitalisation accelerates the discovery, modelling and performance prediction of membrane-based materials for sustainable separation technologies. *Chapter 4* explores the growth potential of emerging materials for decarbonisation, stressing how digitalisation accelerates their discovery, scalability assessment and integration into net-zero technology pathways. *Chapter 5* explores the latest advancements in materials that are driving progress toward net-zero goals. It covers cutting-edge materials for carbon capture, low-carbon hydrogen production and storage and green ammonia production and conversion. *Chapter 6* identifies key industrial barriers to adopting next-generation carbon capture materials and explores how digital tools can help overcome them through predictive modelling, scalability analysis and process integration. *Chapter 7* reviews the solar energy potential in sub-Saharan Africa and explores how digital tools can guide material selection and deployment for sustainable energy transitions. *Chapter 8* explores how digitalisation enhances the development and integration of Power-to-X technologies, enabling efficient material innovation for converting renewable energy into sustainable fuels and chemicals. *Chapter 9* highlights how digital solutions such as AI, modelling and data platforms accelerate materials innovation and deployment for scalable, efficient net-zero technologies across diverse application areas. *Lastly*, the editors conclude the book by creating visions from previous chapters focusing on key themes and future directions in digital materials innovation.

This book can be read for a comprehensive understanding or used as a reference. As we stand at the intersection of urgent climate action and rapid technological advancement, the digitalisation of materials innovation offers a powerful path forward. The chapters ahead are filled with case studies, digital tool applications and summaries to support practical learning along with sustainable and smart progression toward net-zero goals.

REFERENCES

1. Munir, M. T., Naqvi, M., & Li, B. (2024). A converging path: a decade's reflection on net zero emissions and the circular economy. *Frontiers in Energy Research*, 12. https://doi.org/10.3389/fenrg.2024.1332174
2. Bistline, J., Binsted, M. T., Blanford, G., Boyd, G., Browning, M., Cai, Y., Edmonds, J. A., et al. (2025). Policy implications of net-zero emissions: a multi-model analysis of United States emissions and energy system impacts. *Energy and Climate Change*, 6, Article 100191. https://doi.org/10.1016/j.egycc.2025.100191
3. University of Cambridge Institute for Sustainability Leadership (CISL). (2023). *The role of banks and buyers in supporting SMEs to net zero.* https://www.cisl.cam.ac.uk/resources/publications/financial-innovation-sme-net-zero-transition-role-banks-and-buyers
4. Nesta. (2023). *Net zero and productivity in SMEs: overlaps and evidence needs.* https://www.nesta.org.uk/report/net-zero-and-productivity-in-smes-overlaps-and-evidence-needs/
5. Yu, W., & Zhou, Y. (2024). Digital transformation and net-zero innovation in SMEs: evidence from UK firms. *Small Business Economics.* https://doi.org/10.1007/s11187-025-00998-4
6. Zhang, Y., Kumar, S., & Ocampo, L. (2024). Low-cost digital transformation to support sustainability in SMEs: a case-based study. *Sustainability,* 16(3), 1301. https://doi.org/10.3390/su16031301
7. Himanen, L., Geurts, A., Foster, A. S., & Rinke, P. (2019). Data-driven materials science: status, challenges, and perspectives. *arXiv.* https://arxiv.org/abs/1907.05644
8. Make UK. (2023). *Digitalise to decarbonise: digital technology and the journey to net zero.* https://www.makeuk.org/insights/reports/2023/07/11/digitalise-to-decarbonise-report
9. Understanding Manufacturing Energy and Carbon Footprints. https://www1.eere.energy.gov/manufacturing/pdfs/understanding_energy_footprints.pdf
10. How to Calculate the Carbon Footprint of Your Products. https://www.rolandberger.com/en/Insights/Publications/How-to-calculate-the-carbon-footprint-of-your-products.html
11. Manufacturing Energy and Carbon Footprint Definitions and Assumptions. https://www.energy.gov/eere/amo/articles/manufacturing-energy-and-carbon-footprint-definitions-and-assumptions-october
12. Gonzalez-Varona, J. M., Espina-Romero, A., Fernández-Portillo, A., & Martínez-Covarrubias, J. L. (2024). Development of organizational competence for digital transformation in SMEs: a maturity model approach. *arXiv.* https://arxiv.org/abs/2406.01615

Section I

Materials for Decarbonization

2 Amine-Based Materials for CO_2 Capture

Applications, Challenges, and Prospects in Decarbonization

Usman Ali and Muhammad Shahbaz

2.1 INTRODUCTION

The industrial revolution has brought the comforts of modern life through the extensive use of fossil-based materials, which emit significant CO_2, increasing its concentration from 280 ppm in the pre-industrial era to 417 ppm by 2022 [1]. Additionally, the implications of climate change, along with regulations and sustainable development goals, have compelled the world not only to reduce CO_2 emissions but also to focus on its capture, storage and utilization (CCSU) to address these challenges [2,3]. A study reported by the Copenhagen Climate Centre shows that in 2020, greenhouse gas (GHG) emissions were equivalent to 54.4 gigatonnes of CO_2 ($GtCO_2$-eq), and they are expected to reach 60 $GtCO_2$-eq in the future [4]. The world has also signed the Paris Agreement, which aims to limit the rise in global temperatures to below 1.5°C compared to pre-industrial levels. This goal was reaffirmed during climate change conferences, which emphasized the need to reduce GHG emissions to 15 $GtCO_2$-eq [5,6]. In this context, many initiatives, such as negative emissions technologies and net-zero-based technologies, have been introduced, pushing countries to set targets for achieving decarbonization across all sectors, including industry, transport, and lifestyle. These targets vary, with some countries aiming for net zero by 2050, 2060, or 2070 [6]. In this domain, many technologies have been developed, with amine-based solvents and materials at the forefront of decarbonization systems, particularly in CCUS technologies, due to their maturity, current commercial use, and ongoing evolution for future applications [7]. According to available data, there are 19 large-scale physical facilities for CCSU, with 9 based on amine materials, capturing 11 Mt y^{-1} of CO_2 [2]. This large-scale deployment demonstrates the dominance of amine-based materials, which possess physical and chemical properties crucial for industrial-scale applications and environmental solutions.

Growing concerns about emissions have spurred worldwide efforts to transition toward net-zero emission technologies, integrated with digitalization. It is evident that 87 projects have been planned between 2020 and 2030, with amines playing a major

DOI: 10.1201/9781003582304-3

role in combination with other materials and the digitization of current technologies [7,8]. Recent advancements in computing technologies, neural network-based machine learning (ML), and their integration with material development have increased efforts to create frameworks and tools for optimal solutions, such as amine-based materials and their synergy with other materials like ionic liquids (ILs) [7,9]. Amine-based materials offer excellent properties, including high adsorption, reactivity, and versatility in chemical systems. They can also be combined with other materials for gas separation, catalysis, and emerging technologies, with amines classified as primary, secondary, and tertiary based on their organic or inorganic compounds [1,10].

With the advancement of digitalization in science, ML approaches are also progressing in amine-based CO_2 capture technologies, and recent research has explored their potential. ML can play a significant role in optimizing amine solvents for CO_2 capture, particularly in analysing structural properties to develop new amine materials, as reviewed by McDonagh et al. [7]. Another study highlighted the use of cloud computing and spatial analysis for selecting the best amines for CO_2 capture [8]. However, it is clear that limited research has been conducted on the digitization of this field, leaving ample room for further investigation. In CCS technologies, amine-based materials are crucial, and research into the synergy between amines and other materials, such as ILs, is ongoing to enhance capabilities for natural gas (NG) purification and direct air capture (DAC) for net-zero applications. A recent review article emphasized amines' role in CCSU, focusing on their use as CO_2 sorbents and catalyst modifiers. It covered amine-based CO_2 capture principles, adsorption testing, and key adsorbent properties while also exploring nitrogen-modified electrocatalysts for CO_2 utilization, linking lab research to industrial applications to align with net-zero goals.

In CCUS technologies, amine-based materials are crucial, and research into the synergy between amines and other materials, such as ILs, is ongoing to enhance capabilities for NG purification and DAC for net-zero applications. A recent review article emphasized role of amines in CCSU, focusing on their use as CO_2 sorbents and catalyst modifiers. It covered amine-based CO_2 capture principles, adsorption testing, and key adsorbent properties while also exploring nitrogen-modified electrocatalysts for CO_2 utilization, linking lab research to industrial applications to align with net-zero goals [11]. In contrast, Hack et al. [1] reviewed the functionalization of various amine-based materials for CO_2 capture, particularly for DAC systems, a key component of decarbonization in net-zero applications. From the above discussion, it is evident that amine-based materials, due to their commercial viability and emerging potential, are integral to decarbonization efforts and digitization. This study discusses in detail the classification of amines, their chemistry, functionality, degradation, process morphology, and challenges for digitization and net-zero applications.

2.2 CLASSIFICATION OF AMINES

Amines are a group of organic compounds which contain nitrogen atom with a lone pair of electrons. They are structurally similar to ammonia and are distinguished by the number of hydrogen atoms substituted by alkyl or aryl groups that are bonded to the nitrogen atom. The main classifications include the following:

2.2.1 Primary Amines

In methylamine, a primary amine (CH_3NH_2), one hydrogen atom from ammonia (NH_3) is substituted with an aryl or alkyl group. The following are some of the most common forms of primary amines:

- Methylamine (CH_3NH_2) is the most basic form of primary amines, where one of the hydrogen atoms in ammonia is replaced with a methyl group.
- Ethylamine ($C_2H_5NH_2$) is produced when one of the hydrogen atoms in ammonia is replaced with an ethyl group.
- Aniline ($C_6H_5NH_2$) is a type of an aromatic amine because a phenyl group is bonded to the nitrogen atom in its amine structure.
- Benzylamine ($C_6H_5CH_2NH_2$) has a benzyl group attached to its nitrogen atom.

These types of primary amines can be used for a variety of tasks such as making pharmaceutical drugs and dyes or serving as a precursor for organic compound synthesis [12]. The structure of the primary amines is indicated in Figure 2.1a.

2.2.2 Secondary Amines

Dimethylamine ($(CH_3)_2NH$) is a secondary amine where two hydrogen atoms from ammonia are replaced with alkyl or aryl groups. The following are a few illustrations of secondary amines:

- **Dimethylamine ($(CH_3)_2NH$)**: It is a simple secondary amine that has two methyl groups attached to the nitrogen atom.
- **Diethylamine ($(C_2H_5)_2NH$)**: This compound has two ethyl groups bonded to the nitrogen atom.
- **Diphenylamine ($(C_6H_5)_2NH$)**: This is an aromatic amine with two phenyl groups attached to its nitrogen atom.
- **N-Methylaniline ($C_6H_5NHCH_3$)**: A phenyl group and a methyl group are attached to the nitrogen in this case.

Secondary amines, however, find usage in the manufacture of pharmaceuticals, agrochemicals, and dyes, among other things [12]. The structure of the secondary amines is indicated in Figure 2.1b.

2.2.3 Tertiary Amines

A tertiary amine is one in which all three hydrogen atoms in ammonia are substituted with an aryl or alkyl group. Trimethylamine ($N(CH_3)_3$) is an example of a tertiary amine. Here are some forms of tertiary amines:

- **Trimethylamine ($N(CH_3)_2$)**: This compound has three methyl groups bonded to the nitrogen atom.

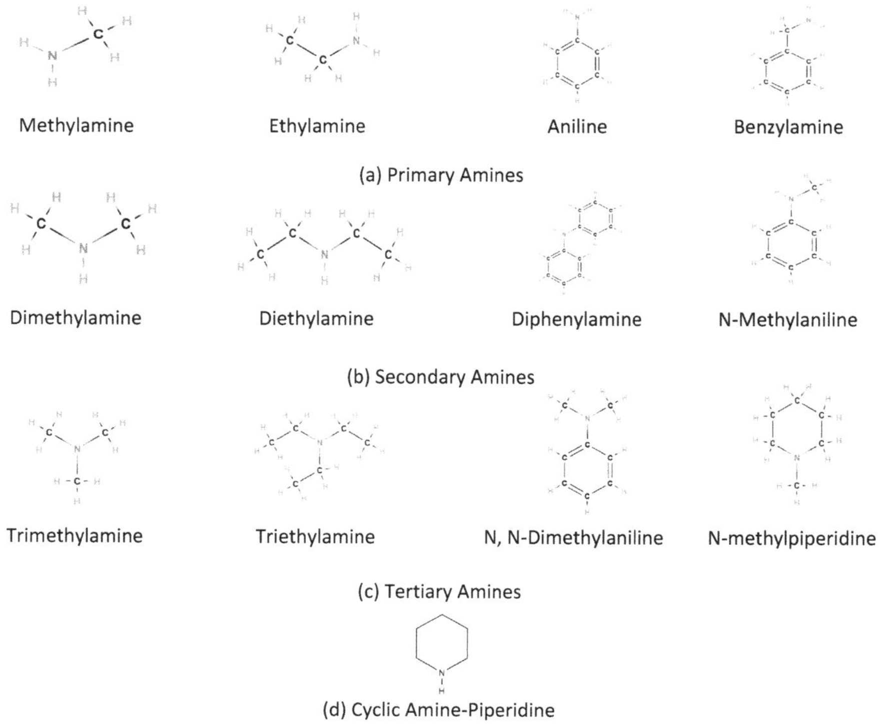

FIGURE 2.1 Structures of amines: (a) primary amines, (b) secondary amines, (c) tertiary amines, and (d) cyclic amines.

- **Triethylamine ($N(C_2H_5)_3$)**: Three ethyl groups are bonded to the nitrogen atom in this amine.
- **N,N-Dimethylaniline ($C_6H_5N(CH_3)_2$)**: This aromatic amine has two methyl groups and one phenyl group bonded to the nitrogen atom.
- **N-Methylpiperidine ($C_5H_{10}NCH_3$)**: A cyclic amine with methyl group bound to the nitrogen atom123.

Tertiary amines have many uses such as catalysts, solvents, or other compounds in organic synthesis [1]. The structure of the tertiary amines is shown in Figure 2.1c.

2.2.4 Cyclic Amines

Cyclic amines are secondary and tertiary amines where nitrogen is a part of a ring structure. Piperidine is a heterocyclic amine, consisting of a six-membered ring containing five methylene bridges ($-CH_2-$) and one amine bridge (–NH–) [13]. The structure of the piperidine is shown in Figure 2.1d. Amines are widely used in the production of pharmaceuticals, dyes, and polymers. They also play a crucial role in biological systems, such as neurotransmitters (e.g., dopamine) and amino acids. Every amine type has its specific property spectrum and is bound to have different chemical reactivity patterns, which is why they find use in different fields of chemistry.

2.3 CHEMISTRY OF AMINES

Amines can be synthesized by using different methods, including the reduction of nitro compounds (R-NO_2) to amines (R-NH_2) and by alkylation of ammonia in which ammonia is reacted with alkyl halides to form primary, secondary, and tertiary amines. The reduction of nitro compounds is generally done through hydrogen in the presence of catalysts, such as platinum (Pt), palladium on carbon (Pd/C), iron (Fe) or zinc (Zn) with acidic medium and tin chloride ($SnCl_2$) in aqueous or alcoholic solutions [14]. The alkylation of ammonia is a nucleophilic type substitution reaction where the nitrogen atom in the ammonia acts as a nucleophile and replaces the halide ion in the alkyl halide. The general mechanism generally involves the nucleophilic attack with a formation of the intermediate, which loses a halide ion forming a primary amine. The primary amine can undergo further alkylation to form secondary or tertiary amines [15].

Amines are basic in nature due to the lone pair of electrons on the nitrogen atom, which can accept any proton (H^+). Lower aliphatic amines show solubility in water due to hydrogen bonding, and it decreases with increasing molecular weight of amines. Monoethanolamine (MEA) is a primary ethanolamine in Figure 2.1. It can associate with H^+ to form an ion $MEAH^+$ and can also react with CO_2 to form a carbamate ion $MEACOO^-$. The basic chemical reactions for the MEA with acid gases (CO_2 and H_2S) are given in Table 2.1.

Diethanolamine (DEA) is a secondary ethanolamine. It can associate with H^+ to form an ion $DEAH^+$ and can also react with CO_2 to form a carbamate ion $DEACOO^-$. The basic chemical reactions between DEA and acid gases (CO_2 and H_2S) can be found in Table 2.2.

Methyldiethanolamine (MDEA) is a tertiary ethanolamine. It can associate with H^+ to form $MDEAH^+$ but cannot react with CO_2 to produce carbamate as primary or secondary ethanolamines can. The basic chemical reactions for the MDEA with acid gases (CO_2 and H_2S) are given in Table 2.3.

The equilibrium constants for Tables 2.1–2.3 can be calculated from the standard Gibbs free energy change. While for the rate-controlled reactions, the power law expressions or the experimental data can be employed. Similar chemistry can be found in the open literature for the other type of amines and/or combination of different amines.

TABLE 2.1
Chemistry and Kinetics of the MEA with CO_2 and H_2S

$MEAH^+ + H_2O \leftrightarrow MEA + H_3O^+$
$MEACOO^- + H_2O \leftrightarrow MEA + HCO_3^-$
$2H_2O \leftrightarrow H_3O^+ + OH^-$
$CO_2 + 2H_2O \leftrightarrow HCO_3^- + H_3O^+$
$HCO_3^- + H_2O \leftrightarrow CO_3^{2-} + H_3O^+$
$H_2S + H_2O \leftrightarrow HS^- + H_3O^+$
$HS^- + H_3O^+ \leftrightarrow S^{-2} + H_3O^+$

TABLE 2.2
Chemistry and Kinetics of the DEA with CO_2 and H_2S

$DEAH^+ + H_2O \leftrightarrow DEA + H_3O^+$
$DEACOO^- + H_2O \leftrightarrow DEA + HCO_3^-$
$2H_2O \leftrightarrow H_3O^+ + OH^-$
$CO_2 + 2H_2O \leftrightarrow HCO_3^- + H_3O^+$
$HCO_3^- + H_2O \leftrightarrow CO_3^{2-} + H_3O^+$
$H_2S + H_2O \leftrightarrow HS^- + H_3O^+$
$HS^- + H_3O^+ \leftrightarrow S^{-2} + H_3O^+$

TABLE 2.3
Chemistry and Kinetics of the MDEA with CO_2 and H_2S

$MDEAH^+ + H_2O \leftrightarrow MDEA + H_3O^+$
$2H_2O \leftrightarrow H_3O^+ + OH^-$
$CO_2 + 2H_2O \leftrightarrow HCO_3^- + H_3O^+$
$HCO_3^- + H_2O \leftrightarrow CO_3^{2-} + H_3O$
$H_2S + H_2O \leftrightarrow HS^- + H_3O^+$
$HS^- + H_3O^+ \leftrightarrow S^{-2} + H_3O^+$

2.3.1 Comparative Potential of Different Amines

Different amines are used for the capture of the CO_2 due to their wide varying properties of the different amines in terms of their efficiencies and properties. MEA, which is primary amine, is the most widely used amine for the CO_2 capture due to its high reactivity with CO_2 and low cost. Though, its regeneration is higher energy intensive, and it also results in corrosion of the process equipment. In comparison to MEA, DEA, which is a secondary amine, is less corrosive and have lower heat of absorption; however, it has lower CO_2 absorption capacity.

In addition, methyldiethanolamine (MDEA), which is a tertiary amine, has higher CO_2 absorption capacity and is less energy intensive, which means its energy requirements are less for regeneration. However, it is less reactive with the CO_2. The mixed amine systems also find application in the CO_2 capture where varying properties of different amines are balanced to enhance the application and achieve the additional efficiencies. Nevertheless, the complexity of the system increases, and optimizing the perfect recipe in terms of mixture ratio will be a bit challenging. Further, the flue gas source defines the amount of the CO_2 in its stream, as NG results in lower CO_2 concentration of about 3–5 vol.%, while the coal and/or biomass combustion generates flue gas of about 13–15 vol.% and industrial flue gas has more than 25 vol. % of the CO_2 in their exhaust streams. Using exhaust gas recirculation to the NG combustion can enhance the CO_2 concentration to about 8–12 vol.%. Hence, each amine has its own set of advantages and disadvantages, and the selection of the specific amine depends on the particular application of the CO_2 capture.

2.4 PROCESS CONSIDERATION FOR AMINES

Amines are widely used for acid gas removal in a process known as amine gas treating or amine scrubbing. This process is essential in industries like oil refineries, petrochemical plants, and NG processing plants to remove hydrogen sulphide (H_2S) and carbon dioxide (CO_2) from gas streams. The recent concern about global warming and climate change have renewed the interest to the development and deployment of the amine-based CO_2 capture system for the treatment of the flue gases to achieve the targets of decarbonization, net zero, and carbon neutral system. The process involves passing the sour gas (containing H_2S and CO_2), flue gas (containing CO_2, N_2, and O_2), and synthesis gas (containing H_2, CO_2, and CO) through an aqueous solution of amines. The amines react with the acid gases to form soluble salts, effectively removing them from the gas stream. Several types of amines are commonly used in this process. MEA is highly effective but can cause corrosion and requires regeneration. DEA is less corrosive than MEA and widely used. MDEA is highly selective for H_2S over CO_2, making it efficient for selective removal. Diisopropanolamine (DIPA) is used in specific applications for its unique properties, while diglycolamine (DGA) is effective for removing both H_2S and CO_2.

2.4.1 Process Morphology

The sour gas is passed through an absorber where it contacts the amine solution. The amine absorbs the acid gases, resulting in a "rich" amine solution. The rich amine solution is then heated in a regenerator (stripper) to release the absorbed gases, regenerating the "lean" amine solution for reuse. The acid gas treatment unit consists of absorbers and stripper whose number varies with the application. The sour/flue/synthesis gas is sent to the booster fan for the pressure increase before it is fed at the bottom of the absorber column which may be a tray or packed column. The gas stream is contacted in a counter-contact manner with the lean amine solvent. The liquid is introduced from the top with the amine solution at lower temperature of particular strength and lean loading depending upon the amine type and regeneration capability. The rich amine solution from the bottom of the absorbers is collected and pumped to the top of the stripper as a single stream. The stream before fed to the stripper is heated through a cross-flow heat exchanger. The acid gases are stripped from the amine solvent and the uncondensed acid gas stream from the condenser is sent to the downstream unit for further processing. The lean amine solution flows down the stripper column and is pumped back for recirculation to the absorber. Further, there is a water wash section at the top of the absorbers to remove entrained droplets of the amine solution in the treated gas before exiting the absorber columns. The generic process flow diagram is shown in Figure 2.2.

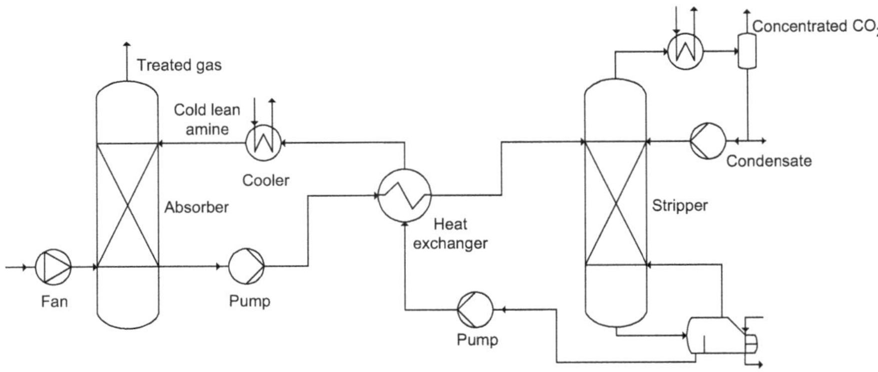

FIGURE 2.2 A typical flow sheet for the acid gas removal [16].

The absorption and regeneration follow the principle of mass transfer which indicates that the absorption is favourable at low temperature and regeneration is favourable at high temperature. Therefore, generally the sour/flue/syn gas temperature needs to be decreased before feeding them to the required temperatures of the absorption process. Further, the flue gas stream upstream process defines the impurities which it will hold. For example, the flue gas from the coal power plant generally contains SO_x and NO_x. Also, the flue gas from the gas-fired power plant will contain the NO_x. These additional components are needed to be removed before the gas stream interaction with the amine solvent. As these components may lead to the formation of the irreversible heat-stable salts due to chemical interaction of these components with the amine as a solvent which is not desirable. It will also result in extra consumption of the solvent inventory and the additional reboiler duty will be consumed. Further, these heat-stable salts may deposit on the internals of the absorber and regeneration columns creating frequent maintenance cost. Depending upon the upstream process type, the gas stream may undergo prior treatments, such as flue gas desulphurization (FGD) for the SO_x removal, selective catalytic reduction unit for the NO_x removal, electrostatic precipitator for the fly ash or soot removal, sulphur polisher for last traces of the SO_x which might have slipped from the FGD, and/or direct contact cooler to decrease the temperature as required by the absorption process [17]. The readers are referred to the literature for different process flow sheets and their details for different types of amines and/or their mixture [12,18].

2.5 DESIGN OF ACID GAS TREATING PROCESS

The design of the gas–liquid contractor is essential to estimate the absorber and stripper design parameters in terms of the diameter and height of the column. An important parameter in this regard is the liquid-to-gas (L/G) ratio, which describes the allowable liquid flow rate to the gas stream flow rate, number of equilibrium stages in both absorber and stripper, and specific energy consumption (SEC) which is generally estimated through the saturated steam flow rate in the reboiler of the stripper [17]. A general rule of thumb from Colburn [19] is that the optimal values of the m(G/L) lie in between 0.7 and 0.8. It depends on the operating and equilibrium line curvatures. Material balances and fluid flow dynamics both can be used for determining the minimum liquid flow rate. The column cross-sectional area and diameter can be determined from the subjected gas flow rate. The number of stages is estimated based on equilibrium data and operating data which will be updated with mass transfer correlations depending on the separation type, whether absorption or regeneration process. Hence, the height of the columns can be estimated. The SEC can be estimated through the overall mechanical and energy balance upon the system. Energy is required for the stripping of the CO_2 from the rich solvent in the reboiler of the stripper. Additionally, cooling utilities are required for the condenser of the stripper and lean solvent cooler. Further, the power is required for the pumps for the lean and rich solvent and for the pumps for the reflux recycle and pump arounds in the absorber, if needed.

The pressure drops which can be tolerated within the column and the approach to maximum capacity are the two parameters which are often used to establish the column diameter for a given gas flow rate and liquid flow rate. Although packed columns are typically constructed within 70%–80% of the flooding velocity, the approach to maximum capacity can range from 70% to 86% of the flooding velocity [20]. The column diameter can be indicated by the equation [20].

$$D = \sqrt{\frac{4G}{\pi U_s}},$$

where G is the gas flow rate and U_s is the superficial velocity which is a function of packing factor and densities of liquid and gas streams involved. Further, the capacity factor is function of the pressure drop per unit height of the packing.

The equivalent to the theoretical plates can be estimated through the following equation [20]:

$$\text{HETP} = \frac{\ln \lambda}{\lambda - 1}\left[\text{HTU}_G + \lambda \text{HTU}_L\right],$$

where λ is the stripping factor function of slope of equilibrium line and L/G ratio and HTU is the height of transfer unit for gas or liquid phases which are a function of superficial velocity for gas and liquid phases, local mass transfer coefficients for the gas and liquid phases, and the effective interfacial area per unit volume of the packed section.

Designing an amine gas treating process is more an art than a science. Several key factors are considered to ensure process effectiveness, efficiency, and integrity. The design dimensions of the absorber and regeneration column are crucial for the proper contact of liquid and gas phases and therefore, the reactivity and absorption of the CO_2 from flue gas stream in the liquid solvent. The concentration, loading, and L/G ratio of the amine used need to be optimized for the better absorption capacity and lower energy requirement for the rich amine.

In terms of the operating condition, both absorption and regeneration are sensitive to the operating temperature and pressure. Absorption is favourable at lower temperature and higher pressure, while stripping is favourable at higher temperature and lower pressure. This fundamental dependency of the temperature and pressure defines the process operating conditions for both absorber and stripper. Similarly, the type and dimensions of the packing material define the efficiency contact between flue gas stream and the solvent and reduce the energy consumption significantly in the stripper. To improve the process performance of the overall system, the advanced control and monitoring of the CO_2 loading, its strength, solvent degradation, and energy consumption across the whole system is generally optimized and implementing the process modifications and alterations, such as heat integration and column intercooling for the reduction of the energy consumption.

2.6 DEGRADATION OF AMINES

The degradation of amines, especially in the context of carbon capture and storage (CCS) technologies, is a significant area of study. Amines are commonly used in CCS to absorb CO_2 from flue gases, but they can degrade over time due to various factors. There are different ways through which degradation of amines may happen, including the following.

2.6.1 Oxidative Degradation

This occurs in the presence of oxygen and can lead to the formation of harmful by-products such as nitrosamines. Oxygen present in the gas stream would cause significant deterioration in terms of the corrosion and overall separation efficiency requiring more solvent and increased specific energy requirements in the reboiler of the regeneration system. Further, the oxygen present in the flue gas may also lead to non-mixing of the conjugate solvents [21].

2.6.2 Thermal Degradation

High temperatures can cause amines to break down, resulting in the loss of solvent and the formation of degradation products. Amines such as MEA and DEA can degrade at high temperatures. For instance, MEA starts to degrade significantly above 177°C. Presence of the acid gases in the gas stream accelerates the thermal degradation process [22]. The degradation of the amines results in the formation of the heat-stable salts as discussed in the previous section. Further, the thermal degradation results in the formation of the ammonia, and other organic products, including formic acid, acetic acid, and oxalic acid. These components may accumulate in the solvent and may escape along the treated gas stream resulting in the environment issues, or create problems in the downstream processes.

2.6.3 Impacts of Degradation

2.6.3.1 Solvent Loss

Degradation reduces the efficiency of the amine solvent, requiring more frequent replacement. It will result in the increased operational cost of the system.

2.6.3.2 Equipment Corrosion

The by-products of degradation can be corrosive, leading to damage to the equipment used in the CCS process. Further, it will require frequent replacement and escalate the maintenance cost. Further, the accumulation of the degradation products can reduce the efficiency of the acid gas removal process.

2.6.3.3 Environmental and Health Concerns

Some degradation products, such as nitrosamines and nitramines, are potentially harmful to human health and the environment.

2.6.4 Mitigation Strategies

2.6.4.1 Improved Solvent Formulations

Developing amines that are more resistant to degradation can help reduce the frequency of solvent replacement. Pretreatment of the gas stream can help to avoid the formation of degradation products.

2.6.4.2 Operational Adjustments

Optimizing the operating conditions, such as temperature and oxygen levels, can minimize degradation. Keeping the operating temperature below the critical threshold which can help minimize the degradation and its products. Regular monitoring of solvent quality and equipment condition can help detect and address degradation issues early.

Mitigating the amine degradation during the CO_2 capture process in the acid gas treating process is crucial for the efficient operation, environment well-being, and process sustainability. Using corrosion and oxidative inhibitors can help protect the equipment from corrosion and reduce the formation of the amine degradation products which is more in the presence of the oxygen and other impurities in the process streams. Furthermore, the improved solvent formulations, in terms of the blended amines and new amine formulations, are more stable and resistant to solvent degradation at a range of operation conditions. Operational optimization in terms of the optimizing the temperature control can help in the minimization of degradation products, along with the process modifications in terms of the feed gas pretreatment to reduce the content of NO_x and SO_x in the gas stream and installing the scrubbers and other technologies for the environmental control of the degradation products being released into the environment. The thermal degradation can be avoided by the proper control of the heat flow to the process and efficient integration of heat with the upstream process, such as steam turbine heat exchangers, bleed steam streams, and/or external heaters. Furthermore, the advanced monitoring of the absorber stack and replenishing the spent solvent with the fresh one will help in making the process more sustainable and affordable.

2.7 HYBRID SOLVENT WITH AMINE-BASED MATERIALS

The new regulations, as well as the development of different materials and processes to enhance the use of hybrid solvents, which combine chemical and physical properties that improve the gas separation efficiency and selectivity, are gaining importance in CCS technologies such as in NG processing and DAC. These solvents are made by a chemical solvent (such as an amine solution) and a physical solvent (such as an absorption liquid or membrane system). These hybrid solvents offer many advantages, such as improved selectivity for pollutants CO_2, H_2S, etc., reduced absorption of other NG components, greater solvent stability, and lower degradation. The physical component is responsible for a longer solvent lifespan, lowering the need for replacement and the environmental impact [23]. Many studies have documented about the use of hybrid amine mixtures to improve CO_2 absorption. In a study by Mandal et al. [23], 2-amino-2-methyl-1-propanol (AMP)-based solutions exhibited higher CO_2 absorption rates than MEA-MDEA mixtures. Wang et al. [24] investigated four MEA-based mixtures (MEA+DIPA, MEA+TEA, MEA+AMP, and MEA+PZ) and discovered that MEA+Piperazine (PZ) had the best mass transfer coefficient, although MEA+AMP had a higher absorption efficiency. In a recent study examined five amine classes (MEA, DEA, AMP, MDEA, and PZ) and interesting blends, including 10 wt.% MEA/20 wt.% PZ and 10 wt.% AMP/20 wt.% PZ, which outperformed regular 30 wt.% MEA solutions in CO_2 capture [25]. Etoumi et al. [26] suggest that piperazine-containing blends can effectively capture CO_2 after combustion. Dash and Bandyopadhyay [27] discovered that combining MDEA with PZ led to higher CO_2 solubility and decreased regeneration energy requirements. Barzagli et al. [28] found that AMP-based blends with alcohols and glycols successfully removed over 90% of CO_2 due to poor carbamate stability. However, the high price of alcohol remains a worry. Gunasekaran et al. [29] examined corrosion in amine blends and found that AMP/PZ and MDEA/PZ were less corrosive than MEA-based systems. Research on amine–alcohol combinations seeks to enhance CO_2 capture efficiency and lower operational costs. Future studies should focus on balancing absorption efficiency, solvent stability, and economic feasibility to promote sustainability in NG processing and CCS.

2.7.1 Optimizing Hybrid Solvents

Hybrid solvents, which mix chemical and physical characteristics, have been intensively studied in process systems engineering to improve gas separation efficiency while overcoming solvent limitations. Amine-based absorption is appropriate for low-CO_2 partial pressure sources, but solvent degradation and energy-intensive regeneration remain challenges. To maximize CO_2 removal and limit degradation, primary and secondary amine blends are tuned for faster reaction rates, better CO_2 loading, and fewer solvent losses. Studies have focused on optimizing hybrid solvent regeneration, reducing energy use, and enhancing solvent management practices. Murtaza et al. [30] observed that MDEA+DGA blends significantly reduced reboiler duty, while Etoumi et al. [26] reported that MDEA/PZ improved CO_2 absorption but increased energy demand. Abotaleb et al. [31] studied that only primary amines and their blends met the NG requirements, with MDEA/PZ (20/10 wt.%) providing the optimum balance of efficiency and energy savings.

Exergy and economic studies have enhanced hybrid solvent usage. Samsami et al. [32] improved an acid gas removal facility, resulting in a 23% reduction in CO_2

emissions and lower steam costs. Jamekhorshid et al. [33] identified absorbers and regenerators as the main sources of energy loss. In another study, it was noticed that hybrid solvents are most effective at CO_2 concentrations between 30.4% and 59.3% [34]. Some researchers have improved absorption efficiency by mixing DGA and piperazine [35,36]. Process systems engineering is driving improvements in hybrid solvent technology by increasing thermodynamics, kinetics, and mass transfer, resulting in more sustainable, cost-effective NG processing options. Future research will concentrate on improving solvent formulations, regeneration techniques, and environmental impact mitigation strategies.

2.8 DIGITIZATION AND MACHINE LEARNING OF SOLVENT

The digitization and integration of ML are significantly transforming process industries. This transformation is also directly influencing the manufacturing and synthesis of materials, particularly in the optimization of material selection, synthesis parameters, and the achievement of desired material properties [37]. Artificial intelligence (AI) is playing a crucial role in organic chemistry, driving innovation not only at the parameter optimization level but also at the molecular design level. Rizvi et al. [38] reviewed the application of AI and ML in organic chemistry, highlighting their use in both supervised and unsupervised learning for data selection and synthesis design. In another article, the progress of AI in solvent screening is explored, along with the challenges of applying mathematical models to detect patterns and reveal relationships within large and complex datasets, especially in the context of identifying green solvents [39]. Priya et al. [40] reviewed a range of studies focused on the application of AI in CO_2 capture technologies, emphasizing its role in optimizing CO_2 capture and storage processes. AI algorithms can significantly enhance the efficiency of these processes by optimizing critical parameters such as temperature, pressure, flow rates, and chemical reactions. Research also suggests that AI-assisted atmospheric CO_2 capture, supported by large-scale monitoring systems, is technically feasible [41]. Additionally, AI enables rapid prediction of CO_2 thermodynamic properties in solutions, thereby streamlining process analysis [40]. These advancements hold promise for applications in pre-combustion CO_2 capture, offering potential improvements in carbon sequestration technologies [42]. In the domain of absorption-based CO_2 capture—where solubility of the absorbent plays a key role—ML has proven useful in predicting physicochemical and thermodynamic properties such as viscosity, density, and solubility. This includes both conventional sorbents like amines and emerging ones like ILs. The prediction of additional thermodynamic properties, including viscosity and density, using ML techniques has also been studied [43]. Hybrid solvents, such as mixtures of ILs with various amines, have shown high potential for enhanced CO_2 absorption. Experimental investigations are ongoing, as solubility remains a critical factor in these systems. Ali et al. [44] developed deep learning models to predict CO_2 solubility in ILs, utilizing a dataset of 10,116 solubility measurements across 164 types of ILs under diverse temperature and pressure conditions.

2.9 CHALLENGES AND OPPORTUNITIES

2.9.1 Opportunities and Challenges

The use of amine-based materials will increase over time with their application in decarbonization and net-zero initiatives. The integration of digitalization brings many changes and opportunities.

- The thermal degradation of amine materials due to CO_2, moisture, and high temperatures reduces their capability to capture CO_2 and lowers efficiency. The durability and functionality of amines at high temperatures remain a challenge. The formation of hybrid materials can address many issues, discussed earlier with different amine materials [31]. The use of ILs with amines, metal–organic frameworks, nanomaterials, and composite materials can enhance efficiency and improve durability.
- An important issue is the production of amine-based materials, which is an energy-intensive process that directly affects the economic feasibility of their application. Therefore, the development of efficient carbon desorption or regeneration methods is necessary to improve the process [10].
- Another significant challenge is the innovation and development of amine-based materials with high efficiency but scaling them up to industrial levels remains costly due to manufacturing limitations. It is essential to develop process models with techno-economic studies to expand their use in various applications.
- Other mitigation technologies, such as DAC, have emerged in the last decade, where amine-based materials have shown promise. However, using amine-based materials for CO_2 capture from air or flue gas presents challenges, as these sources have different CO_2 concentrations compared to conventional amine applications [11].
- The use of ML and AI can help optimize material selection, synthesis processes, process efficiency, and integration into new and existing industrial applications. However, this requires extensive data, computing skills, and specialized industry expert for adoption [7].
- AI and ML can also accelerate the discovery of novel amine-based materials with improved stability, lower energy requirements, and higher CO_2 capture efficiency. Additionally, predictive modelling can optimize amine formulations and reaction conditions to enhance performance [8].
- The use of digital twins and smart monitoring, along with data collection, will help optimize real-world processes and reduce energy penalties. Furthermore, advancements in sensor development and automation can enhance process efficiency.
- Developing bio-based or recyclable amine materials can make carbon capture more sustainable. Innovative solvent-free or low-energy synthesis routes can also reduce costs and environmental impact.

REFERENCES

1. J. Hack, N. Maeda, and D. M. Meier, "Review on CO_2 capture using amine-functionalized materials," *ACS Omega,* vol. 7, no. 44, pp. 39520–39530, 2022. https://doi.org/10.1021/acsomega.2c03385
2. H. Yamada, "Amine-based capture of CO_2 for utilization and storage," *Polymer Journal,* vol. 53, no. 1, pp. 93–102, 2021. https://doi.org/10.1038/s41428-020-00400-y
3. M. Shahbaz et al., "A comprehensive review of biomass based thermochemical conversion technologies integrated with CO_2 capture and utilisation within BECCS networks," *Resources, Conservation and Recycling,* vol. 173, p. 105734, 2021. https://doi.org/10.1016/j.resconrec.2021.105734
4. P. United Nations Environment, *Emissions Gap Report 2019.* UN, 2022. : The Closing Window — Climate crisis calls for rapid transformation of societies. Nairobi. https://www.unep.org/emissions-gap-report-2022
5. M. Shahbaz, M. Alherbawi, E. C. Okonkwo, and T. Al-Ansari, "Evaluating negative emission technologies in a circular carbon economy: a holistic evaluation of direct air capture, bioenergy carbon capture and storage and biochar," *Journal of Cleaner Production,* vol. 466, p. 142800, 2024. https://doi.org/10.1016/j.jclepro.2024.142800
6. P. N. S. B. Nair, R. R. Tan, and D. C. Y. Foo, "Extended graphical approach for the implementation of energy-consuming negative emission technologies," *Renewable and Sustainable Energy Reviews,* vol. 158, p. 112082, 2022.
7. J. L. McDonagh et al., "Machine guided discovery of novel carbon capture solvents," *arXiv preprint arXiv:2303.14223,* 2023.
8. J. McDonagh et al., "Chemical space analysis and property prediction for carbon capture amine molecules." *Digital Discovery*, vol. 3, no. 3, pp. 528–543, 2023.
9. M. C. Davis, W. J. M. Kort-Kamp, I. Matanovic, P. Zelenay, and E. F. Holby, "Design of amine-functionalized materials for direct air capture using integrated high-throughput calculations and machine learning," *arXiv preprint arXiv:2410.13982,* 2024.
10. Z. Liang, K. Fu, R. Idem, and P. Tontiwachwuthikul, "Review on current advances, future challenges and consideration issues for post-combustion CO_2 capture using amine-based absorbents," *Chinese Journal of Chemical Engineering,* vol. 24, no. 2, pp. 278–288, 2016. https://doi.org/10.1016/j.cjche.2015.06.013
11. L. B. Hamdy, C. Goel, J. A. Rudd, A. R. Barron, and E. Andreoli, "The application of amine-based materials for carbon capture and utilisation: an overarching view," *Materials Advances,* vol. 2, no. 18, pp. 5843–5880, 2021.
12. A. L. Kohl and R. Nielsen, *Gas Purification.* Elsevier, 1997.
13. S. A. Lawrence, *Amines: Synthesis, Properties and Applications.* Cambridge University Press, 2004.
14. A. Ricci and L. Bernardi, *Methodologies in Amine Synthesis: Challenges and Applications.* Wiley-VCH GmbH, 2021.
15. R. Chen, T. Zhang, Y. Guo, J. Wang, J. Wei, and Q. Yu, "Recent advances in simultaneous removal of SO_2 and NO_x from exhaust gases: removal process, mechanism and kinetics," *Chemical Engineering Journal,* vol. 420, p. 127588, 2021.
16. U. Ali, K. J. Hughes, D. B. Ingham, L. Ma, and M. Pourkashanian, "Effect of the CO_2 enhancement on the performance of a micro gas turbine with a pilot-scale CO_2 capture plant," *Chemical Engineering Research and Design,* vol. 117, pp. 11–23, 2017. https://doi.org/10.1016/j.cherd.2016.10.001
17. J. Wilcox, *Carbon Capture.* Springer Science & Business Media, 2012.
18. A. I. Papadopoulos and P. Seferlis, *Process Systems and Materials for CO_2 Capture: Modelling, Design, Control and Integration.* John Wiley & Sons, 2017.

19. A. P. Colburn, *The Simplified Calculation of Diffusional Processes: General Consideration of Two-Film Resistances*. American Society of Mechanical Engineers, 1939.
20. E. O. Agbonghae, K. J. Hughes, D. B. Ingham, L. Ma, and M. Pourkashanian, "Optimal process design of commercial-scale amine-based CO_2 capture plants," *Industrial & Engineering Chemistry Research,* vol. 53, no. 38, pp. 14815–14829, 2014. https://doi.org/10.1021/ie5023767
21. C. Gouedard, D. Picq, F. Launay, and P. L. Carrette, "Amine degradation in CO_2 capture. I. A review," *International Journal of Greenhouse Gas Control,* vol. 10, pp. 244–270, 2012.
22. G. T. Rochelle, "Thermal degradation of amines for CO_2 capture," *Current Opinion in Chemical Engineering,* vol. 1, no. 2, pp. 183–190, 2012.
23. B. P. Mandal, M. Guha, A. K. Biswas, and S. S. Bandyopadhyay, "Removal of carbon dioxide by absorption in mixed amines: modelling of absorption in aqueous MDEA/MEA and AMP/MEA solutions," *Chemical Engineering Science,* vol. 56, no. 21–22, pp. 6217–6224, 2001.
24. W. Wang, Q. Li, Y. Liu, and B. Chen, "Ionic liquid-aqueous solution ultrasonic-assisted extraction of three kinds of alkaloids from Phellodendron amurense Rupr and optimize conditions use response surface," *Ultrasonics Sonochemistry,* vol. 24, pp. 13–18, 2015.
25. A. Adeosun, N. El Hadri, E. Goetheer, and M. R. Abu-Zahra, "Absorption of CO_2 by amine blends solution: an experimental evaluation," *International Journal of Engineering and Science,* vol. 3, no. 9, pp. 12–23, 2013.
26. A. Etoumi, M. Alhanash, M. Almabrouk, and M. Emtir, "Performance improvement of gas sweetening units by using a blend of MDEA/PZ," *Chemical Engineering Transactions,* vol. 86, pp. 1057–1062, 2021.
27. S. K. Dash and S. S. Bandyopadhyay, "Studies on the effect of addition of piperazine and sulfolane into aqueous solution of N-methyldiethanolamine for CO_2 capture and VLE modelling using eNRTL equation," *International Journal of Greenhouse Gas Control,* vol. 44, pp. 227–237, 2016.
28. F. Barzagli, F. Mani, and M. Peruzzini, "Efficient CO_2 absorption and low temperature desorption with non-aqueous solvents based on 2-amino-2-methyl-1-propanol (AMP)," *International Journal of Greenhouse Gas Control,* vol. 16, pp. 217–223, 2013.
29. P. Gunasekaran, A. Veawab, and A. Aroonwilas, "Corrosivity of single and blended amines in CO_2 capture process," *Energy Procedia,* vol. 37, pp. 2094–2099, 2013.
30. A. Murtaza, K. Qureshi, and I. N. Unar, "Energy optimization for amine gas sweetening process by mixed amines using simulations," *Engineering Science and Technology, an International Journal,* vol. 3, pp. 35–40, 2019.
31. A. Abotaleb, M. H. El-Naas, and A. Amhamed, "Enhancing gas loading and reducing energy consumption in acid gas removal systems: a simulation study based on real NGL plant data," *Journal of Natural Gas Science and Engineering,* vol. 55, pp. 565–574, 2018.
32. S. Samsami, M. Mohamadi, M. H. Sarrafzadeh, E. R. Rene, and M. Firoozbahr, "Recent advances in the treatment of dye-containing wastewater from textile industries: overview and perspectives," *Process Safety and Environmental Protection,* vol. 143, pp. 138–163, 2020/11// 2020. https://doi.org/10.1016/J.PSEP.2020.05.034
33. A. Jamekhorshid, Z. K. Davani, A. Salehi, and A. Khosravi, "Gas sweetening simulation and its optimization by two typical amine solutions: an industrial case study in Persian Gulf region," *Natural Gas Industry B,* vol. 8, no. 3, pp. 309–316, 2021.
34. B. Zhang, X. Zhao, X. Wu, M. Han, C. H. Guan, and S. Song, "Consumption-based accounting of global anthropogenic CH_4 emissions," *Earth's Future,* vol. 6, no. 9, pp. 1349–1363, 2018.
35. C. J. Ejeh, G. P.-H. Mawufemor, and A. Ransford, "Sulfolane and di-iso-propanol lean amine blend operating temperature and pressure effect to natural gas sweetening using process simulation," *SN Applied Sciences,* vol. 2, pp. 1–11, 2020.

36. R. M. El-Maghraby, A. A. Salah, and A. M. Shoaib, "Carbon dioxide capturing from natural gas using di-glycol amine and piperazine–a new solvent mixture," *International Journal of Recent Technology and Engineering (IJRTE),* vol. 8, no. 4, pp. 11378–11383, 2019.
37. M. E. Mondejar et al., "Digitalization to achieve sustainable development goals: steps towards a Smart Green Planet," *Science of The Total Environment,* vol. 794, p. 148539, 2021. https://doi.org/10.1016/j.scitotenv.2021.148539
38. R. S. Aal E Ali, J. Meng, M. E. I. Khan, and X. Jiang, "Machine learning advancements in organic synthesis: a focused exploration of artificial intelligence applications in chemistry," *Artificial Intelligence Chemistry,* vol. 2, no. 1, p. 100049, 2024. https://doi.org/10.1016/j.aichem.2024.100049
39. J. P. Edaugal, D. Zhang, D. Liu, V.-A. Glezakou, and N. Sun, "Solvent screening for separation processes using machine learning and high-throughput technologies," *Chem & Bio Engineering,* vol. 2, no. 4, pp. 210–228, 2025. https://doi.org/10.1021/cbe.4c00170
40. A. K. Priya, B. Devarajan, A. Alagumalai, and H. Song, "Artificial intelligence enabled carbon capture: a review," *Science of the Total Environment,* vol. 886, p. 163913, 2023. https://doi.org/10.1016/j.scitotenv.2023.163913
41. S. E. M. Elhenawy, M. Khraisheh, F. AlMomani, and G. Walker, "Metal-organic frameworks as a platform for CO_2 capture and chemical processes: adsorption, membrane separation, catalytic-conversion, and electrochemical reduction of CO_2," *Catalysts,* vol. 10, no. 11, p. 1293, 2020.
42. C. Altintas, O. F. Altundal, S. Keskin, and R. Yildirim, "Machine learning meets with metal organic frameworks for gas storage and separation," *Journal of Chemical Information and Modeling,* vol. 61, no. 5, pp. 2131–2146, 2021.
43. M. Rahimi, S. M. Moosavi, B. Smit, and T. A. Hatton, "Toward smart carbon capture with machine learning," *Cell Reports Physical Science,* vol. 2, no. 4, p. 100396, 2021. https://doi.org/10.1016/j.xcrp.2021.100396
44. M. Ali et al., "Prediction of CO_2 solubility in Ionic liquids for CO_2 capture using deep learning models," *Scientific Reports,* vol. 14, no. 1, p. 14730, 2024. https://doi.org/10.1038/s41598-024-65499-y

3 Digitalization and Computational Developments of Membrane Materials for Net-Zero Emissions

Rizwan Nasir, Eleonora Ricci,
Nor Naimah Rosyadah Ahmad,
Siti Zulaikha Mohd Yusof, and Khuram Maqsood

3.1 INTRODUCTION

Atmospheric carbon dioxide (76%) acts as a dominant (76%) greenhouse gas, followed by methane (16%) and nitrous oxide (2%) [1]. CO_2 plays a major contributor to global warming, posing significant risks to human well-being, ecosystems [2], and the economy [1,3]. Without intervention, the growing levels of carbon emissions are predicted to raise the global temperature, which could lead to a 3°C–4°C average temperature increase worldwide by the end of the century [1]. As CO_2 emissions act as a primary cause of climate change, it is important to minimize the release of heat-trapping greenhouse CO_2 gases into the atmosphere. Membrane technology has gained recognition as an effective method for CO_2 capture in various applications, including natural gas processing, hydrogen purification, flue gas post-treatment, and biogas upgrading [4]. In addition to gas separation processes, membranes have been used in a number of other industries, such as CO_2 capture, wastewater treatment, membrane crystallization, and chemical and pharmaceutical processes [2]. Membranes are currently thought of as a second-generation technology for absorbing and removing CO_2 [5]. For the process and chemical industries, which are essential to combating climate change and advancing environmental sustainability, the transition to net-zero emissions is a critical undertaking [6]. Among net-zero technologies, membrane processes have emerged as a crucial industrial separation technology and are anticipated to be essential to the development of sustainable production systems in the future [2]. Various membranes, such as polymeric, inorganic, and mixed matrix-facilitated transport membranes, and gas–liquid membrane contractors, have been developed for CO_2 separation [7–9]. Several international industrial companies are commercializing polymeric membranes to separate CO_2 from other gases

DOI: 10.1201/9781003582304-4

TABLE 3.1
Industrial Companies Involved in Membrane CO_2 Separation

Industrial Company	Membrane Type	Application
DIVEX GLOBAL	Helipur®	CO_2 from $N_2/Ar/O_2$
AIRRANE Co, Ltd.	Polysulfone hollow fiber	CO_2 from $CH_4/H_2/N_2$
IGS	Generon®	CO_2/H_2
PBI Performance Products, Inc.	Celazole® (Polybenzimidazole hollow fiber)	CO_2/H_2
Compact Membranes Systems, Inc.	Fluoropolymer	CO_2/CH_4; CO_2/N_2
CYNARA	Cellulose triacetate	CO_2 from natural gas
UOP – A Honeywell Company	–	CO_2/H_2
GRACE, SEPAREX	Cellulose acetate	–
MEDAL	Polyimide-based	–
AIR Liquide S.A.	–	CO_2 from biogas/natural gas
MTR	Perfluoro-polymers	–
EVONIK Industries AG	Sepuran®	CO_2 from biogas

Source: Adapted from A. Iulianelli and E. Drioli, [10] with Permission Copyright (2020) from Elsevier.

(Table 3.1) [10]. Membrane separation is part of an innovation of advanced separation methods, following phase processes like extraction, adsorption, and absorption, as well as traditional thermal techniques like distillation and evaporation [2,11]. The membrane materials play a huge role in achieving the net-zero target and mitigating climate change. They can also withstand harsh industrial environments, which ensure long-term operational stability.

3.1.1 Overview of Digitalization in Membrane Innovation

Significant efforts have been invested in the fabrication of membranes and optimizing the potential of the membrane processes, to lower energy consumption, enhance the separation of capacity, and prolong the membrane lifespan [12,13]. Apart from operational and synthesis conditions, the performance of the membrane in separation and purification can also be influenced by its chemical, functional, and structural properties [14]. Researchers are eager to have a deeper understanding of these separation mechanisms of membranes and attempt to develop reliable machine learning (ML) algorithms that can forecast the functionality and characteristics of these membranes. In recent years, hybrid models based on artificial intelligence (AI) have emerged from the growth of empirical and phenomenological modeling. Computational demands can be minimized by developing simpler models that capture essential features [12,15]. To address the drawbacks of conventional mechanistic models and make it possible for future advancements, AI-based approaches have been proposed as helpful tools for gaining insightful into filtration and membrane separation processes [16]. A crucial area of AI known as ML has contributed to the research and development standards for new environmental and energy materials. The process varies depending on the type of membrane (2D materials or polymeric), influencing the choice of ML models for prediction [17]. Molecular modeling methods, such as density functional theory and molecular dynamics simulations, give a significant result for material design and optimization, which offer detailed insights into molecular transport mechanisms to enable rationale material design [18,19]. These techniques enable researchers to forecast the characteristics and effectiveness of the composites, optimize their properties for applications, and model the atomic-level interactions between polymer matrices and nanoparticles [18].

3.1.2 Digital Tools for Membrane Design and Optimization

With the considerable success of ML in several domains of scientific discovery, researchers have been addressing number of open challenges through data-driven ML methods. The utilization of models for optimization and screening reduces the time for experimental synthesis [17]. Improving the design of the hollow-fiber membrane module is necessary to improve the separation performance and increase the energy and resource efficiency of the overall process. This can be done by applying digital tools such as AI innovation, computational fluid dynamics (CFD), and cloud computing, leading to greater predictive analytics and energy efficiency. CFD is known as a valuable tool for the membrane separation process in gaining deeper insights with a broad range of membrane separation [20]. The integration of CFD simulation in the application of membrane distillation (MD) process analysis is extensively used for the development of membrane modules, novel membrane design, and also in analyzing the MD structural features and mechanisms [21]. Apart from that, AI technology has proved to be reliable to overcome the drawbacks of conventional mechanistic models, as it has become a very significant and applicable tool nowadays for anticipating the future employment of membrane separation and filtration process [12]. The advancement and integration of digital tools into membrane science studies will lead to much smarter, more efficient, and more sustainable separation technologies. Table 3.2 summarizes the key digital tools used for membrane design and optimization.

In recent years, ML techniques have shown their unquestionable potential for use in the synthesis, manufacturing, and parameter assessment of polymeric membranes. As a result, they may be a viable prescreening tool to reduce costly trial-and-error experimental techniques. These methods could significantly increase the accuracy of polymer characteristic prediction, make it easier to develop and synthesize polymeric membranes, and eventually develop new, polymers with a wide range of potential uses in membrane-based separation processes. AI has received much attention lately because of its strong prediction performance, affordability, ease of use, adaptability, and excellent capacity to optimize various membrane applications. Numerous scientific studies on membrane separation methods are demonstrating promising results for various technical applications [22]. The use of ML techniques in membrane-based separation technology is based on dataset size and target demand [23]. Distinct AI-based techniques, including artificial neural network, support vector machines, deep learning, and Gaussian process regression, have been extensively used in many predictive mathematical-based models in recent years [24–26]. ML can accelerate the cycle and facilitate the development of a new membrane and membrane

TABLE 3.2
Key Digital Tools Used for Membrane Design and Optimization

Tool	Functionality	Applications in Membrane Science	Key Features	Examples/Tools	References
Density functional theory	Quantum mechanical modeling of electronic structure	Predicting adsorption properties of membrane surfaces	High-accuracy modeling, electronic structure optimization	VASP, Gaussian, Quantum ESPRESSO	[32,33]
Molecular dynamics simulation	Studies atomic-level interactions of membrane materials	Understanding polymer swelling, gas permeability, and membrane aging	Atomistic modeling, force-field parameters, and solvation effects	GROMACS, LAMMPS, NAMD	[34,35]
Membrane computing simulators	Models biological and computational processes using P systems for dynamic simulations of membrane interactions	Synthetic biology, ecological modeling, and process automation	Building, analyzing, and optimizing	P-Lingua, Infobiotics Workbench	[36]
Computer-aided design	Processes simulation for chemical and separation processes	Membrane-integrated process modeling for desalination and gas separation	Thermodynamic modeling, process optimization	Aspen Plus, gPROMS, DWSIM	[35,37]
Computational fluid dynamics	Simulates fluid flow through membranes and modules	Optimization of membrane module designs for better performance	Multiphase flow modeling, turbulence prediction	ANSYS Fluent, COMSOL Multiphysics	[38,39]
Finite element analysis	Solves complex fluid dynamics and mechanical stress analysis	Structural integrity analysis of membranes under different pressures	Mesh-based simulations, stress–strain modeling	COMSOL Multiphysics, ANSYS	[40,41]
Artificial intelligence and machine learning	Predicts membrane performance and material properties	Membrane material discovery, rejection efficiency predictions, predicting separation performance, and membrane fouling behavior	Regression/classification models, deep learning frameworks, Nonlinear modeling	MATLAB, Python (Scikit-learn, TensorFlow, PyTorch), Chemoinformatics Tools	[17,42]

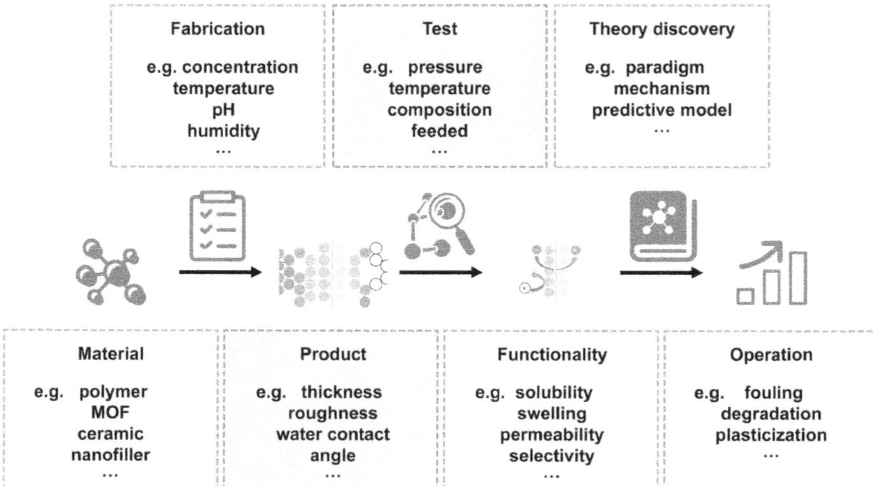

FIGURE 3.1 Membrane discovery cycle (highlighted areas correspond to the stages to which AI can be applied). (Adapted from Yin et al. [27] under the Creative Commons license.)

process, as shown in Figure 3.1 [27]. Numerous developments in membrane science and technology have emerged as a result of the increased interest in using AI and especially ML [28,29]. For example, researchers' growing understanding of the pros and cons of integrating AI methods with CFD has improved the precision and effectiveness of turbulence problem solutions [30,31]. Accurate ML predictions of gas selectivity and permeability in polymeric membranes can significantly accelerate the screening process for optimal candidates. Similarly, ML can predict key desalination performance metrics in water treatment applications, including salt rejection and water flux, enhancing membrane evaluation and selection.

3.1.3 Sustainable Practices in Membrane Technology

The development that meets the needs of the present without compromising the ability of future generations to meet their own needs is known as sustainable development [43]. It is essential to consider the incorporation of environmental, social, economic, and technical factors of the desired activities [44]. According to the literature, sustainability criteria for the development of membrane-based technologies must include technical factors such as treatment efficiency, hybrid techniques, ease of implementation, process stability, and health and safety risks; economic include operating and maintenance costs, and initial investments; environmental includes solid waste generation, release of chemical substances, CO_2 emission, water reuse potential, and possibility to recover by-products, and social includes public acceptance, odor and noise impact, and visual impact [45]. Furthermore, other sustainable alternatives include biodegradable polymers, the integration of green solvents, and the addition of additives. The current literature highlights the growing emphasis on innovative and sustainable manufacturing techniques, emphasizing the adoption of eco-friendly materials, advanced methodologies, and efforts to minimize waste and cost. Future studies should generally take a comprehensive approach, as shown in Table 3.3, which addresses material selection and fabrication process enhancement, and includes biodegradable elements in developing sustainable gas separation membranes [46]. Recycling membranes and utilizing recycled material for manufacturing is a potential approach. Depending on physicochemical properties, the end-of-life membrane could be reutilized for similar, upgraded, and downgraded operations, thus extending the membrane lifespan while mitigating the environmental impact that occurred due to their disposal and new membrane preparation for similar purposes. Likewise, using recycled waste such as polystyrene, terephthalate, polyethylene, tire rubber, polyvinyl chloride, keratin, and cellulose and their derivatives for fabricating the membranes can significantly enhance environmental sustainability [47].

TABLE 3.3
Pillars of Sustainable Gas Separation Membrane

Materials	**Sustainable materials for gas separation membranes should be :** • Withstand harsh chemical environments and elevated temperatures • Enhanced mechanical properties • Achieved a balance between selectivity and permeability • Designed to resist fouling and degradation • Controlled pore size and distribution can be precisely controlled (tailored pore structure) • Adapted to incorporate chemical functional groups
Techniques	**Sustainable manufacturing techniques focus on reducing environmental impact, conserving energy, while maintaining or improving the performance of the membranes:** • Using bio-based or biodegradable materials in 3D printing/additive manufacturing • Solvent-free 3D printing • Cross-linking strategies that do not rely on hazardous chemicals • Using water as a solvent and bio-based polyelectrolytes in layer-by-layer assembly
Green manufacturing	**The concept of green membrane manufacturing encompasses several key aspects:** • Biodegradable, recyclable, or derived from renewable sources to minimize environmental impact • Minimizes the use of harmful chemicals in the membrane fabrication process • Engineered to operate with reduced energy consumption • Implementing recycling and waste reduction measures in manufacturing facilities • Sustainable fabrication, using less energy, using renewable energy sources, and generating fewer emissions

Source: Adapted from Alkandari and Castro-Dominguez [46], under Creative Commons Attribution License (CC BY).

3.2 COMPUTATIONAL MODELING AND AI METHODS FOR MEMBRANE MATERIALS SELECTION

Scientists and engineers have traditionally relied on domain expertise and empirical tests to select suitable materials for membrane separation and have identified progressively better structures by trial-and-error process. This workflow hinders lab-to-market times, due to the vast number of possible material combinations and the complexity of their property relationships, which cannot be evaluated extensively via experiments. As their accuracy continues to improve over the years, computational models are evolving from being an after-the-fact addition to experimental campaigns, used to provide explanations of results and interpolations between tested conditions, to be used as predictive tools for known and unknown materials at a scale that is orders of magnitude greater than could be tested in the laboratory, and the modeled structures are successfully passing experimental validation tests in the search for net-zero materials [48]. Physics-based models, such as molecular simulations [49] and data-driven methods, such as ML techniques [50] are accelerating materials discovery as well as aiding in process optimization and control, allowing membrane technology to achieve better energy efficiency and sustainability [16,27,51].

3.2.1 Mechanistic Insights from Molecular Modeling

Molecular modeling methods provide an extremely high level of detail into materials' behavior and can be applied to explain the microscopic molecular mechanisms that shape the macroscopic separation of membrane materials. In particular, MD and Monte Carlo simulations have been extensively used to predict the solubility, diffusivity, and permeability of gaseous and liquid species in membranes for carbon capture [49,52–54], hydrogen purification [55,56], fuel cells [57,58], water treatment [59], and pervaporation [60]. In addition to property prediction, their strength lies in providing insight into transport mechanisms [61] and analysis of the microstructure of membrane materials and its effect on the separation performance [62]. They have contributed to the understanding of the effects of functionalization [63], as well as interface compatibility in the case of nanocomposite membrane materials [52,64,65], demonstrating how tailored organic–inorganic interfaces can significantly enhance membrane stability and separation efficiency [66].

3.2.2 High-Throughput Screening and Data-Driven Methods

Data-driven methods are revolutionizing numerous science and engineering applications. By discovering patterns and relationships between the data, ML algorithms are increasingly effective tools for materials discovery and selection in membrane applications for net zero as well, as discussed in recent review papers focusing on the application of ML methods to materials for gas separations [50,67,68], water purification and wastewater treatment [42,69,70], and membranes for fuel cells [71,72] and electrolyzers [73,74]. Some of the challenges encountered in the application of data-driven methods in membrane science are the relatively small amount of data available, for ML standards, and the need to explore efficiently a very vast design space. Despite these limitations, studies have successfully demonstrated how reliable models could be obtained even with limited experimental data available [75], also leveraging data augmentation or imputation techniques [76].

Data-driven methods are often applied in the context of high-throughput screening (HTS). HTS allows to evaluate the properties of a large number of materials to search for promising candidates, either by brute force combinatorial search (which typically requires extensive computing power) or through search algorithms. In recent years, the application of active learning techniques has been on the rise for the efficient exploration of vast materials design spaces. Active learning is a ML method in which the algorithm selectively queries an external system (such as a molecular simulation, or an experimental facility) to obtain information for specific data points, that are expected to be most informative. To efficiently build structure–property relationships, active learning focuses on identifying and learning from the most uncertain or representative samples, optimizing the learning process, and reducing the required data. This technique is particularly useful when obtaining data is expensive or time-consuming, as it enhances model performance with fewer samples [23]. A typical workflow is exemplified in Figure 3.2 [77], from a study that targeted the discovery of polyimides of intrinsic microporosity for CO_2/CH_4, CO_2/N_2 and O_2/N_2 separation.

Another valuable approach in the case of limited data is transfer learning, a ML method where a model developed initially for a specific task is subsequently adapted for a different, related task. This approach leverages knowledge gained from one domain to improve learning in another, often requiring less data and training time. For membrane materials discovery, transfer learning can be applied by pre-training models on datasets for separations that are well studied (more data available) and then fine-tuning the model to make predictions on a novel separation. This strategy was applied, for example [78], to transfer knowledge from the CO_2/CH_4 separation to the CO_2/N_2 for the screening of mixed matrix membranes. Oftentimes, researchers are interested in solving an inverse-design problem, i.e., finding materials that satisfy a certain performance threshold. Genetic algorithms are ML-constrained optimization methods used to find optimal solutions by iteratively evolving a population through selection, crossover, and mutation. This method has also found application in the case of membrane materials discovery [79], and allowed the identification of new polymer membranes for CO_2/N_2 and CO_2/O_2 separations.

Transformer architectures are on the limelight for their pivotal role in the success of large language models, and they are being tested in materials science applications as well. Kuenneth and Ramprasad developed a "chemical language model", polyBERT [80], based on a transformer architecture. It was trained on the SMILES strings of 100 million hypothetical polymers and coupled to a multitask neural network to predict around 30 different polymer properties, including six pure gas permeabilities. A challenge associated to data-driven methods is their limited explainability, what is known as their "black-box" character. The development of explainable ML method is a rapidly evolving area of research, and methods such as SHAP (SHapley Additive exPlanations), feature importance analysis, or partial dependence plots have been employed also in membrane materials discovery to enhance model interpretability and provide insights into the underlying structure–property relationships [78,81,82].

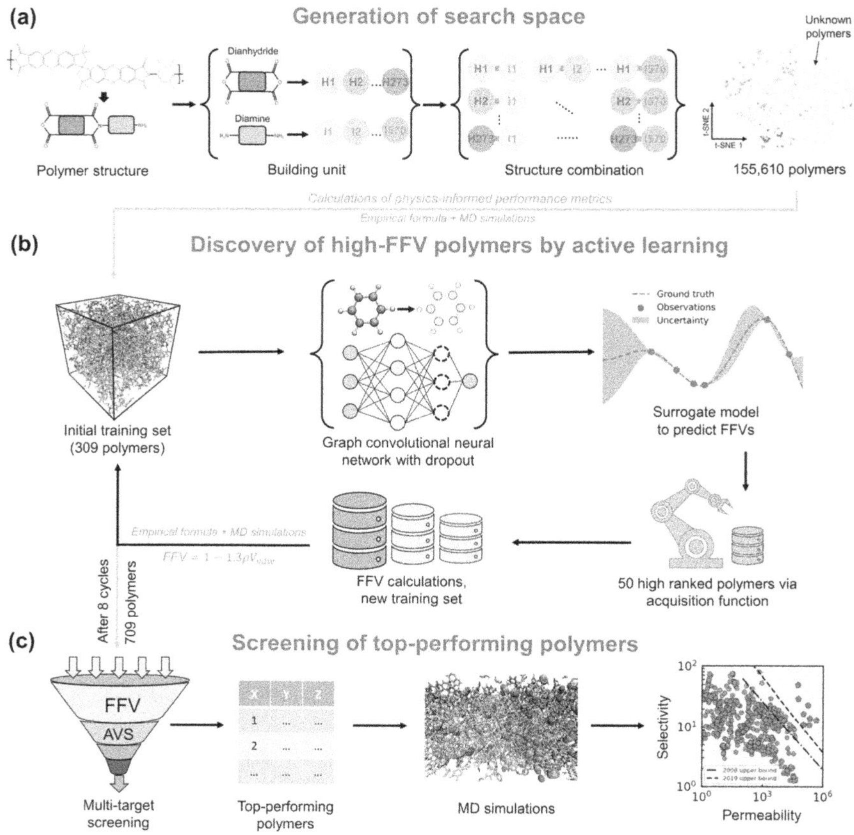

FIGURE 3.2 Active learning high-throughput modeling workflow. (a) A large search space is identified, (b) then active learning is applied to sample from it and construct a dataset of molecular simulation results. (c) Top-performing materials are identified by multi-target screening, and MD simulations evaluate their separation performance. (Reproduced from [77] under CC-BY 4.0 license, © 2024 The Authors. Advanced Functional Materials published by Wiley-VCH GmbH.)

3.3 MEMBRANE PROCESS OPTIMIZATION AND SIMULATION

These days, various membrane applications are being developed and optimized with AI, enabling the most established ones to be updated and improved to operate more effectively [83]. The literature reports on the modeling and simulation of membrane processes for different applications. These processes can be simulated using various programming languages and simulation software tools. Many studies focus on modeling the effectiveness of membrane separation procedures [84]. The optimization of a membrane system using models can be divided into two parts: modeling the membrane itself and programming the optimization problem [85]. In the 1980s, the membranes were used for processes with high sour gas concentrations. It was later shown that while membranes are cost-effective for removing the majority of sour gas, they cannot reach natural gas (retentate) when the concentration of sour gas is less than 2%. For the bulk separation, a hybrid system (amine + membrane) was introduced. The amine mechanism simultaneously lowers the retentate's sour concentration to the desired level. Most researchers agree that membranes can separate materials more effectively in bulk. Most industrial applications can be divided into two main categories for separation procedures. Membrane performance is better than other separation techniques, provided that the target component (TC) concentration in the feed is high. On the contrary, when the feed has a low TC concentration, poor membrane performance is commonly observed due to a single membrane stage that could not attain high purity in the permeate line and low concentration in the retentate line. Higher membrane areas and sufficient pressure will ensure the best design meets recovery and purity requirements. The performance of membranes under such low-concentration feed conditions is questionable, because such configurations result in greater CAPEX (because of the high membrane area) and OPEX (because of the expensive compression costs). Combining membrane systems with several steps set up in series or parallel to achieve higher quality has been the solution for such situations [86].

3.4 DIGITAL TWINS AND MEMBRANE TECHNOLOGY

The most recent development in physical system models is represented by digital twins (DT), which continuously couple the modeling and physical domains of an experimental setup [87,88]. The fundamental concept behind the DT was to develop a system that combines virtual and physical components, utilizing the advantages of each to the system's advantage. Notably, the singled out to be the feedback (update) between the virtual and the physical domains is a critical component of a DT [89] enabling (a) predictive control of the physical system [90] and (b) the update of the virtual states of the system based on data [91]. Product data are collected, saved, analyzed, and used for present and upcoming products. This approach essentially makes it possible to apply an up-to-date data-driven strategy for tracking, overseeing, and enhancing a product across the course of its life cycle [89].

The physical performance of entities is being enhanced through DT, which is composed of a physical system, a virtual model, and the data network between them, using the computational tools provided by the virtual model [92]. The parameters, boundary conditions, and dynamics of the virtual model are developed using real-time data from the physical entity, producing an accurate representation of the real entities being modeled [93]. The steps of a DT process are shown in Figure 3.3 [94]. While DT technology has seen limited application in membrane-based systems, particularly in water and gas separation processes, its transformative potential aligns seamlessly with the goals of net-zero applications.

A decision support system (DSS) was developed by Rooij et al. to assess membrane restoration approaches. The DSS is operated by a DT, which provides a computerized visualization of membrane element wear (degradation) and restoration. The DT is based on a mathematical model that depicts a reverse osmosis vessel as a multicomponent system where pieces can be swapped or changed and wear-states of individual elements (components) are measured. The findings indicate that the modeled and

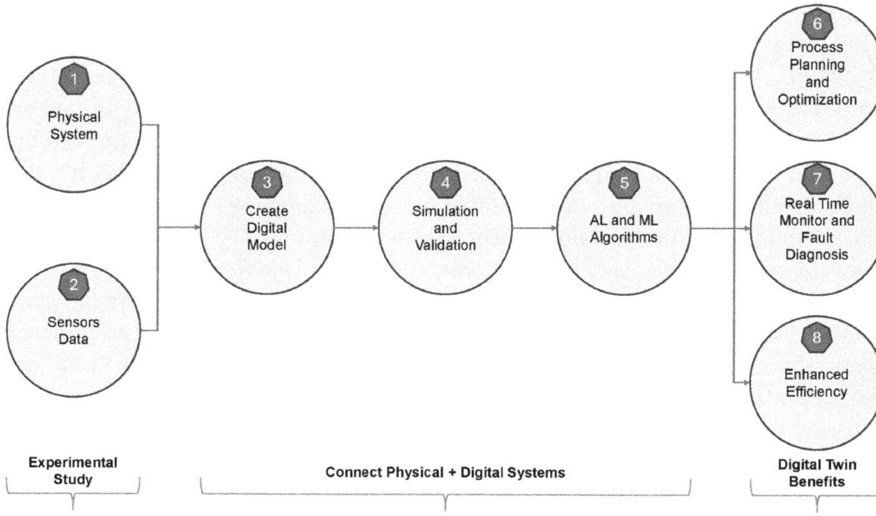

FIGURE 3.3 Illustration of digital twin process. (Adapted with permission from Ghenai et al. [94], Copyright (2022) from Elsevier.)

observed wear-states fit one another well [95]. Lin et al. used experimental data to develop a 3D DT representation of a hollow-fiber membrane. A multiscale simulation was also used to successfully examine the gas filtration process of ultrafine particles utilizing a hollow-fiber membrane. The results showed greater accuracy than the results of the experimental measurements [96]. Using statistical techniques to process data, Menezes et al. presented a methodology. Then, they used it for the first time as a web application to enable online and real-time monitoring of spiral wound membranes for gas separation membrane processes. For future stages of simulation and parameter estimation, reliable data were provided [97].

3.5 LIFE CYCLE ASSESSMENT AND ENVIRONMENTAL IMPACT ANALYSIS OF MEMBRANE MATERIALS AND PROCESSES

Understanding the environmental impacts of membrane separation is essential to ensuring sustainable practices and can help optimize energy consumption. The life cycle assessment (LCA) approach has been developed to evaluate the environmental footprints of any technology, including membrane separation processes. This assessment spans from the very beginning of raw materials to the residue degradation aspects, illustrated by the "cradle-to-grave" scenario [98]. LCA is a reliable and efficient tool for evaluating the sustainability of new technologies, since it can effectively predict the cause-and-effect relationships within the environment [99]. Figure 3.4 illustrates the LCA framework, which involves several steps, including the identification of goals and scope of the project, the creation of the life cycle inventory (LCI), the life cycle impact assessment (LCIA), and the interpretation of the results based on the LCI and LCIA. Improvements can be made to the other processes based on the insights gained from each stage [98].

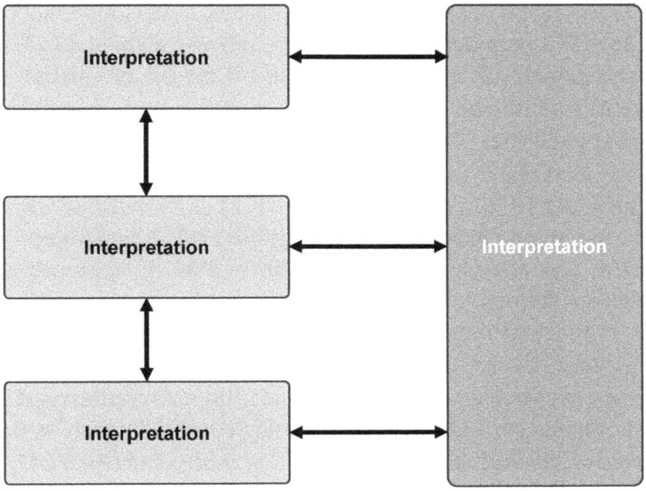

FIGURE 3.4 Illustration of LCA framework based on ISO 14040/44. (Adapted with permission from Fu et al. [98], © 2024 Elsevier B.V.)

Currently, LCA has been used widely to assess various strategies and technologies for net-zero applications such as waste management [100,101], renewable energy systems [102–104], and CO_2 capture and storage [105]. However, there are still limited studies on LCA for membrane-based processes compared to other technologies in CO_2 capture such as amine-based absorption. Troy et al. [106] evaluated the real greenhouse gas emissions and other environmental implications of novel membrane separation methods via LCA approach. It was found that membrane technology showed superior outcomes compared to the traditional CO_2 capture methods for most impact categories, including environmental aspects. Additionally, they concluded that factors such as membrane area and lifespan directly affect the material demand. However, the performance of polymeric membranes requires improvement to compete with amine washing for post-combustion CO_2 capture application. In another work, Giordano et al. [107] assessed the environmental impacts of the membrane process in post-combustion CO_2 capture using the hybrid method for the LCI, where both economic and physical data inputs were combined. Their simulation results demonstrated that the gas permeation properties of polymers significantly impacted the life cycle emissions of the membrane separation for the capture systems. Additionally, it was proposed that the membrane selective layer thickness and material were closely associated with the membrane area and net power consumption. Other environmental consequences could be amplified by increasing membrane area due to the greater emissions in the membrane production stage [108]. In order to achieve the best possible balance between environmental consequences and performance, it was suggested to employ a membrane-based CO_2 capture process by combining membranes with high selectivity and permeability [98,107].

The techno-economic study was conducted and examined the environmental impacts of an integrated catalytic conversion process and CO_2 capture in producing synthetic natural gas (SNG). A polymer electrolyte membrane was used to simulate the integrated plant involving the electrolysis process. It was found that CO_2 emission was reduced by 71% compared to the reference process without a CO_2 capture unit. The greater the amount of captured CO_2 utilized for the production of SNG, the more it displaces fossil-based natural gas, hence diminishing fossil resource depletion. It was suggested that plant energy consumption could be further reduced by considering the membrane or thermal swing adsorption technology in improving the CO_2 capture unit [109]. Fan et al. [99] used LCA to evaluate the environmental impacts of membrane-based direct air capture (DAC) integrated with CO_2 utilization (m-DAC-U). They concluded that the m-DAC-U approach can not only be carbon-negative through carbon capture and utilization but can also drastically reduce its environmental impact by including recycling and reuse concepts during the end-of-life phase. Even so, the economic feasibility aspect needs to be improved by reducing the expensive capital cost. Fu et al. [98] analyzed several LCA studies based on post-combustion CO_2 capture technologies, which focus on the membrane separation process. Overall, it was suggested that post-combustion CO_2 capture systems utilizing membrane technology are anticipated to mitigate global warming potential, provided that the membrane system does not require excessive electricity consumption. Nonetheless, the membrane production and sustainability aspect of membrane materials require improvement to further mitigate additional environmental footprints. Additionally, future works on LCA for membrane separation can be improved using AI, specifically for several aspects, including data collection and processing. For instance, the LCA accuracy can be improved by reducing the data input uncertainty, which can be achieved through real-time monitoring of membrane performance and emissions. Various datasets from experimental and industrial operations can be analyzed using AI to refine the LCI step. In terms of the automated LCA framework, ML algorithms can improve the LCIA by identifying relationships between life cycle stages and their environmental effects.

3.6 FUTURE PROSPECTS IN DIGITALIZED MEMBRANE TECHNOLOGY

The future of membrane materials and manufacturing will depend on the innovation, scalability, and sustainability of new approaches, the stability, the feasibility of delivering rapid prototypes, and collaborative efforts with risk-taking visionary industries [110]. The integration of computational modeling and AI methods is transforming the materials selection landscape. However, challenges remain to realize the full potential of these methods. These include high computational requirements, the need for high-quality comprehensive datasets, and the complexity of integrating physics-based and data-driven models to achieve computationally efficient screening with interpretable results [50,67,81]. Although significant progress has been made in the in silico predictive design of membrane materials, the field is still in its beginning, and it is essential to seek experimental validation whenever possible, while also leveraging explainable ML methods to enhance the reliability of results and promote broader adoption of computational approaches in materials discovery workflows. ML-powered membrane technology provides environmentally friendly options for energy production, gas separation, and water treatment. Membranes combined with ML provide the potential for a cleaner and more sustainable future by addressing water scarcity, lowering greenhouse gas emissions, and improving energy production efficiency. The entire potential of membrane technology will be further unlocked by ongoing research and development in this area, supporting the worldwide effort to preserve the environment and promote sustainable development [42].

ACKNOWLEDGMENT

All authors are thankful to their affiliated institutions for providing academic and support facilities for the completion of this work.

REFERENCES

1. K. R. Shivanna, "Climate change and its impact on biodiversity and human welfare," *Proceedings of the Indian National Science Academy,* vol. 88, pp. 160–171, 2022.
2. A. Kamolov, Z. Turakulov, S. Rejabov, G. Díaz-Sainz, L. Gómez-Coma, A. Norkobilov, et al., "Decarbonization of power and industrial sectors: the role of membrane processes," *Membranes,* vol. 13, p. 130, 2023.
3. L. J. R. Nunes, "The rising threat of atmospheric CO_2: a review on the causes, impacts, and mitigation strategies," *Environments,* vol. 10, p. 66, 2023.
4. Y. Jia, K. Wong, C. Z. Liang, J. Wu, T.-S. Chung, and S. Zhang, "Recent development of membranes for carbon capture: from materials to asymmetric membranes," *Progress in Materials Science,* vol. 146, p. 101324, 2024.
5. R. Hou, C. Fong, B. D. Freeman, M. R. Hill, and Z. Xie, "Current status and advances in membrane technology for carbon capture," *Separation and Purification Technology,* vol. 300, p. 121863, 2022.
6. P. Glavič, Z. N. Pintarič, H. Levičnik, V. Dragojlović, and M. Bogataj, "Transitioning towards net-zero emissions in chemical and process industries: a holistic perspective," *Processes,* vol. 11, p. 2647, 2023.
7. R. Khalilpour, K. Mumford, H. Zhai, A. Abbas, G. Stevens, and E. S. Rubin, "Membrane-based carbon capture from flue gas: a review," *Journal of Cleaner Production,* vol. 103, pp. 286–300, 2015.
8. B. Belaissaoui and E. Favre, "Membrane separation processes for post-combustion carbon dioxide capture: state of the art and critical overview," *Oil & Gas Science and Technology–Revue d'IFP Energies Nouvelles,* vol. 69, pp. 1005–1020, 2014.
9. W. Y. Hong, "A techno-economic review on carbon capture, utilisation and storage systems for achieving a net-zero CO_2 emissions future," *Carbon Capture Science & Technology,* vol. 3, p. 100044, 2022.
10. A. Iulianelli and E. Drioli, "Membrane engineering: latest advancements in gas separation and pre-treatment processes, petrochemical industry and refinery, and future perspectives in emerging applications," *Fuel Processing Technology,* vol. 206, p. 106464, 2020.
11. P. Markewitz, W. Kuckshinrichs, W. Leitner, J. Linssen, P. Zapp, R. Bongartz, et al., "Worldwide innovations in the development of carbon capture technologies and the utilization of CO_2," *Energy & Environmental Science,* vol. 5, pp. 7281–7305, 2012.
12. M. Jafari, C. Tzirtzipi, and B. Castro-Dominguez, "Applications of artificial intelligence for membrane separation: a review," *Journal of Water Process Engineering,* vol. 68, p. 106532, 2024.

13. C. Niu, X. Li, R. Dai, and Z. Wang, "Artificial intelligence-incorporated membrane fouling prediction for membrane-based processes in the past 20 years: a critical review," *Water Research,* vol. 216, p. 118299, 2022.
14. R. Dangayach, N. Jeong, E. Demirel, N. Uzal, V. Fung, and Y. Chen, "Machine learning-aided inverse design and discovery of novel polymeric materials for membrane separation," *Environmental Science & Technology,* vol. 59, pp. 993–1012, 2025.
15. M. J. Regufe, V. V. Santana, A. F. P. Ferreira, A. M. Ribeiro, J. M. Loureiro, and I. B. R. Nogueira, "A hybrid modeling framework for membrane separation processes: application to lithium-ion recovery from batteries," *Processes,* vol. 9, p. 1939, 2021.
16. N. D. Viet, D. Jang, Y. Yoon, and A. Jang, "Enhancement of membrane system performance using artificial intelligence technologies for sustainable water and wastewater treatment: a critical review," *Critical Reviews in Environmental Science and Technology,* vol. 52, pp. 3689–3719, 2022.
17. Z. Cao, O. B. Farimani, J. Ock, and A. B. Farimani, "Machine learning in membrane design: from property prediction to AI-guided optimization," *Nano Letters,* vol. 24, pp. 2953–2960, 2024.
18. I. Rezić and M. S. Škoc, "Computational methodologies in synthesis, preparation and application of antimicrobial polymers, biomolecules, and nanocomposites," *Polymers,* vol. 16, p. 2320, 2024.
19. S. B. Levy and B. Marshall, "Antibacterial resistance worldwide: causes, challenges and responses," *Nature Medicine,* vol. 10, pp. S122–S129, 2004.
20. B. Lukitsch, P. Ecker, M. Elenkov, C. Janeczek, B. Haddadi, C. Jordan, et al., "Computation of global and local mass transfer in hollow fiber membrane modules," *Sustainability,* vol. 12, p. 2207, 2020.
21. L. Chen and B. Wu, "Research progress in computational fluid dynamics simulations of membrane distillation processes: a review," *Membranes,* vol. 11, p. 513, 2021.
22. S. Yuan, H. Ajam, Z. A. B. Sinnah, F. M. A. Altalbawy, S. A. Abdul Ameer, A. Husain, et al., "The roles of artificial intelligence techniques for increasing the prediction performance of important parameters and their optimization in membrane processes: a systematic review," *Ecotoxicology and Environmental Safety,* vol. 260, p. 115066, 2023.
23. A. Tayyebi, A. S. Alshami, X. Yu, and E. Kolodka, "Can machine learning methods guide gas separation membranes fabrication?," *Journal of Membrane Science Letters,* vol. 2, p. 100033, 2022.
24. W. S. Noble, "What is a support vector machine?," *Nature Biotechnology,* vol. 24, pp. 1565–1567, 2006.
25. B. Wang and T. Chen, "Gaussian process regression with multiple response variables," *Chemometrics and Intelligent Laboratory Systems,* vol. 142, pp. 159–165, 2015.
26. J. Zuo, H. Lv, D. Zhou, Q. Xue, L. Jin, W. Zhou, et al., "Deep learning based prognostic framework towards proton exchange membrane fuel cell for automotive application," *Applied Energy,* vol. 281, p. 115937, 2021.
27. H. Yin, M. Xu, Z. Luo, X. Bi, J. Li, S. Zhang, et al., "Machine learning for membrane design and discovery," *Green Energy & Environment,* vol. 9, pp. 54–70, 2024.
28. Y. Cao, A. T. Nakhjiri, S. M. Sarkar, and M. Ghadiri, "Time-dependent numerical investigation of 3-hydroxypropionic acid extraction using a microporous membrane contactor," *The European Physical Journal Plus,* vol. 137, p. 494, 2022.
29. J. Jawad, A. H. Hawari, and S. J. Zaidi, "Artificial neural network modeling of wastewater treatment and desalination using membrane processes: a review," *Chemical Engineering Journal,* vol. 419, p. 129540, 2021.
30. M. Babanezhad, A. Masoumian, A. T. Nakhjiri, A. Marjani, and S. Shirazian, "Influence of number of membership functions on prediction of membrane systems using adaptive network based fuzzy inference system (ANFIS)," *Scientific Reports,* vol. 10, p. 16110, 2020.

31. H. Azimi, S. Shabanlou, I. Ebtehaj, H. Bonakdari, and S. Kardar, "Combination of computational fluid dynamics, adaptive neuro-fuzzy inference system, and genetic algorithm for predicting discharge coefficient of rectangular side orifices," *Journal of Irrigation and Drainage Engineering,* vol. 143, p. 04017015, 2017.
32. G. Zhang, M. Pérez-Jiménez, A. Riscos-Núñez, S. Verlan, S. Konur, T. Hinze, et al., *Membrane Computing Models:* Implementations. Springer, Singapore, 2021.
33. S. Raghavan and K. Chandrasekaran, "Tools and simulators for membrane computing-a literature review," in Gong, M., Pan, L., Song, T., Zhang, G. (eds.), *Bio-Inspired Computing – Theories and Applications.* Communications in Computer and Information Science, vol 681Springer, Singapore, 2016, pp. 249–277.
34. W. J. Koros and C. Zhang, "Materials for next-generation molecularly selective synthetic membranes," *Nature Materials,* vol. 16, pp. 289–297, 2017.
35. P. T. Mitkowski, W. Szaferski, A. Nędzarek, and M. Sales-Cruz, "Design of membrane systems," *Physical Sciences Reviews,* vol. 8, pp. 4029–4051, 2023.
36. N. Bhardwaj, R. V. Stahelin, R. E. Langlois, W. Cho, and H. Lu, "Structural bioinformatics prediction of membrane-binding proteins," *Journal of Molecular Biology,* vol. 359, pp. 486–495, 2006.
37. M. Li, "Modeling, simulation, and optimization of membrane processes," *Separations,* vol. 10, p. 303, 2023.
38. R. Baghel, S. Kalla, S. Upadhyaya, S. P. Chaurasia, and K. Singh, "CFD modeling of vacuum membrane distillation for removal of Naphthol blue black dye from aqueous solution using COMSOL multiphysics," *Chemical Engineering Research and Design,* vol. 158, pp. 77–88, 2020.
39. Q. Zhu, W. Cheng, J. Yang, H. Sun, W. Liu, and H. Li, "Optimizing efficiency of proton exchange membrane electrolyzer system based on multiphysics model and differential evolution strategy," *Journal of Power Sources,* vol. 621, p. 235270, 2024.
40. Z. Lin, Y. Wang, Z. Fu, D. Zhong, X. Du, and J. Zhang, "Optimization of mechanical properties and finite element simulation analysis of hollow fiber membranes in gas-liquid membrane separation technology," *Desalination,* vol. 599, p. 118456, 2025.
41. G. Liu, R. Guo, K. Zhao, and R. Wang, "Monte Carlo simulation for exploring mechanical properties of porous materials based on scaled boundary finite element method," *Applied Sciences,* vol. 12, no. 2, p. 575, 2022.
42. A. I. Osman, M. Nasr, M. Farghali, S. S. Bakr, A. S. Eltaweil, A. K. Rashwan, et al., "Machine learning for membrane design in energy production, gas separation, and water treatment: a review," *Environmental Chemistry Letters,* vol. 22, pp. 505–560, 2024.
43. G. H. Brundtland and M. Khalid, *Our Common Future.* Oxford University Press, Oxford, GB, 1987.
44. C.-W. Chen, "Guidance on the conceptual design of sustainable product–service systems," *Sustainability,* vol. 10, p. 2452, 2018.
45. M. Kamali, D. P. Suhas, M. E. Costa, I. Capela, and T. M. Aminabhavi, "Sustainability considerations in membrane-based technologies for industrial effluents treatment," *Chemical Engineering Journal,* vol. 368, pp. 474–494, 2019.
46. S. H. Alkandari and B. Castro-Dominguez, "Advanced and sustainable manufacturing methods of polymer-based membranes for gas separation: a review," *Frontiers in Membrane Science and Technology,* vol. 3, p. 1390599, 2024.
47. N. K. Khanzada, R. A. Al-Juboori, M. Khatri, F. E. Ahmed, Y. Ibrahim, and N. Hilal, "Sustainability in membrane technology: membrane recycling and fabrication using recycled waste," *Membranes,* vol. 14, p. 52, 2024.
48. H. Park, X. Yan, R. Zhu, E. A. Huerta, S. Chaudhuri, D. Cooper, et al., "A generative artificial intelligence framework based on a molecular diffusion model for the design of metal-organic frameworks for carbon capture," *Communications Chemistry,* vol. 7, p. 21, 2024.

49. A. Mollahosseini and A. Abdelrasoul, "Molecular dynamics simulation for membrane separation and porous materials: a current state of art review," *Journal of Molecular Graphics and Modelling,* vol. 107, p. 107947, 2021.
50. E. Ricci and M. G. De Angelis, "A perspective on data-driven screening and discovery of polymer membranes for gas separation, from the molecular structure to the industrial performance," *Reviews in Chemical Engineering,* vol. 40, pp. 567–600, 2024.
51. X. Wang, K. R. Hinkle, C. J. Jameson, and S. Murad, "Using molecular simulations to facilitate design and operation of membrane-based and chiral separation processes," *Industrial & Engineering Chemistry Research,* vol. 61, pp. 15554–15566, 2022.
52. S. J. Datta, A. Mayoral, N. Murthy Srivatsa Bettahalli, P. M. Bhatt, M. Karunakaran, I. D. Carja, et al., "Rational design of mixed-matrix metal-organic framework membranes for molecular separations," *Science,* vol. 376, pp. 1080–1087, 2022.
53. H. Mashhadimoslem, M. A. Abdol, P. Karimi, K. Zanganeh, A. Shafeen, A. Elkamel, et al., "Computational and machine learning methods for CO_2 capture using metal–organic frameworks," *ACS nano,* vol. 18, pp. 23842–23875, 2024.
54. H. Demir, G. O. Aksu, H. C. Gulbalkan, and S. Keskin, "MOF membranes for CO_2 capture: past, present and future," *Carbon Capture Science & Technology,* vol. 2, p. 100026, 2022.
55. H. Daglar and S. Keskin, "Combining machine learning and molecular simulations to unlock gas separation potentials of MOF membranes and MOF/polymer MMMs," *ACS Applied Materials & Interfaces,* vol. 14, pp. 32134–32148, 2022.
56. S. Aydin, C. Altintas, and S. Keskin, "High-throughput screening of COF membranes and COF/polymer MMMs for helium separation and hydrogen purification," *ACS Applied Materials & Interfaces,* vol. 14, pp. 21738–21749, 2022.
57. R. Jinnouchi, K. Kudo, K. Kodama, N. Kitano, T. Suzuki, S. Minami, et al., "The role of oxygen-permeable ionomer for polymer electrolyte fuel cells," *Nature Communications,* vol. 12, p. 4956, 2021.
58. J. Cheng, G. He, and F. Zhang, "A mini-review on anion exchange membranes for fuel cell applications: stability issue and addressing strategies," *International Journal of Hydrogen Energy,* vol. 40, pp. 7348–7360, 2015.
59. Q. Xu and J. Jiang, "Molecular simulations of liquid separations in polymer membranes," *Current Opinion in Chemical Engineering,* vol. 28, pp. 66–74, 2020.
60. Q. Sun, H. Ma, L. Wu, J. Ding, L. Wang, and Y. Hu, "Molecular simulation for guiding the design and optimization of mixed matrix membranes (MMMs) in the pervaporation process," *Langmuir,* vol. 39, pp. 5199–5210, 2023.
61. M. Heiranian, R. M. DuChanois, C. L. Ritt, C. Violet, and M. Elimelech, "Molecular simulations to elucidate transport phenomena in polymeric membranes," *Environmental Science & Technology,* vol. 56, pp. 3313–3323, 2022.
62. Y. Liu, N. Li, X. Cui, W. Yan, J. Su, and L. Jin, "A review on the morphology and material properties of the gas separation membrane: molecular simulation," *Membranes,* vol. 12, p. 1274, 2022.
63. J. Yang, Z. Shen, J. He, and Y. Li, "Efficient separation of small organic contaminants in water using functionalized nanoporous graphene membranes: insights from molecular dynamics simulations," *Journal of Membrane Science,* vol. 630, p. 119331, 2021.
64. R. Semino, N. A. Ramsahye, A. Ghoufi, and G. Maurin, "Microscopic model of the metal–organic framework/polymer interface: a first step toward understanding the compatibility in mixed matrix membranes," *ACS applied materials & interfaces,* vol. 8, pp. 809–819, 2016.
65. R. C. Dutta and S. K. Bhatia, "Interfacial engineering of MOF-based mixed matrix membrane through atomistic simulations," *The Journal of Physical Chemistry* C, vol. 124, pp. 594–604, 2019.

66. S.-L. Wu, F. Liu, H.-C. Yang, and S. B. Darling, "Recent progress in molecular engineering to tailor organic–inorganic interfaces in composite membranes," *Molecular Systems Design & Engineering,* vol. 5, pp. 433–444, 2020.
67. G. Ignacz, L. Bader, A. K. Beke, Y. Ghunaim, T. Shastry, H. Vovusha, et al., "Machine learning for the advancement of membrane science and technology: a critical review," *Journal of Membrane Science,* vol. 713, p. 123256, 2025.
68. J. Wang, K. Tian, D. Li, M. Chen, X. Feng, Y. Zhang, et al., "Machine learning in gas separation membrane developing: ready for prime time," *Separation and Purification Technology,* vol. 313, p. 123493, 2023.
69. Q. Xu and J. Jiang, "Recent development in machine learning of polymer membranes for liquid separation," *Molecular Systems Design & Engineering,* vol. 7, pp. 856–872, 2022.
70. P. Dansawad, Y. Li, Y. Li, J. Zhang, S. You, W. Li, et al., "Machine learning toward improving the performance of membrane-based wastewater treatment: a review," *Advanced Membranes,* vol. 3, p. 100072, 2023.
71. R. Ding, S. Zhang, Y. Chen, Z. Rui, K. Hua, Y. Wu, et al., "Application of machine learning in optimizing proton exchange membrane fuel cells: a review," *Energy and AI,* vol. 9, p. 100170, 2022.
72. Z. Rui, R. Ding, K. Hua, X. Duan, X. Li, Y. Wu, et al., "Design of proton exchange membranes with high durability for fuel cells: from the perspective of machine learning," *Journal of Membrane Science,* vol. 683, p. 121831, 2023.
73. M. Batool, O. Sanumi, and J. Jankovic, "Application of artificial intelligence in the materials science, with a special focus on fuel cells and electrolyzers," *Energy and AI,* vol. 18, p. 100424, 2024.
74. R. Ding, Y. Chen, Z. Rui, K. Hua, Y. Wu, X. Li, et al., "Machine learning utilized for the development of proton exchange membrane electrolyzers," *Journal of Power Sources,* vol. 556, p. 232389, 2023.
75. J. W. Barnett, C. R. Bilchak, Y. Wang, B. C. Benicewicz, L. A. Murdock, T. Bereau, et al., "Designing exceptional gas-separation polymer membranes using machine learning," *Science Advances,* vol. 6, p. eaaz4301, 2020.
76. Q. Yuan, M. Longo, A. W. Thornton, N. B. McKeown, B. Comesaña-Gándara, J. C. Jansen, et al., "Imputation of missing gas permeability data for polymer membranes using machine learning," *Journal of Membrane Science,* vol. 627, p. 119207, 2021.
77. M. Wang and J. Jiang, "Accelerating discovery of polyimides with intrinsic microporosity for membrane-based gas separation: synergizing physics-informed performance metrics and active learning," *Advanced Functional Materials,* vol. 34, no. 23, p. 2314683, 2024.
78. J. Guan, T. Huang, W. Liu, F. Feng, S. Japip, J. Li, et al., "Design and prediction of metal organic framework-based mixed matrix membranes for CO_2 capture via machine learning," *Cell Reports Physical Science,* vol. 3, no. 5, p. 100864, 2022.
79. Y. Basdogan, D. R. Pollard, T. Shastry, M. R. Carbone, S. K. Kumar, and Z.-G. Wang, "Machine learning-guided discovery of polymer membranes for CO_2 separation with genetic algorithm," *Journal of Membrane Science,* vol. 712, p. 123169, 2024.
80. C. Kuenneth and R. Ramprasad, "polyBERT: a chemical language model to enable fully machine-driven ultrafast polymer informatics," *Nature Communications,* vol. 14, p. 4099, 2023.
81. J. Yang, L. Tao, J. He, J. R. McCutcheon, and Y. Li, "Machine learning enables interpretable discovery of innovative polymers for gas separation membranes," *Science Advances,* vol. 8, p. eabn9545, 2022.
82. N. Jeong, R. Epsztein, R. Wang, S. Park, S. Lin, and T. Tong, "Exploring the knowledge attained by machine learning on ion transport across polyamide membranes using explainable artificial intelligence," *Environmental Science & Technology,* vol. 57, pp. 17851–17862, 2023.

83. M. T. Gaudio, G. Coppola, L. Zangari, S. Curcio, S. Greco, and S. Chakraborty, "Artificial intelligence-based optimization of industrial membrane processes," *Earth Systems and Environment,* vol. 5, pp. 385–398, 2021.
84. R. Abejón, C. Casado-Coterillo, and A. Garea, "Techno-economic optimization of multistage membrane processes with innovative hollow fiber modules for the production of high-purity CO_2 and CH_4 from different sources," *Industrial & Engineering Chemistry Research,* vol. 61, pp. 8149–8165, 2022.
85. Z. Rizki and M. Ottens, "Model-based optimization approaches for pressure-driven membrane systems," *Separation and Purification Technology,* vol. 315, p. 123682, 2023.
86. A. Alshehri, R. Khalilpour, A. Abbas, and Z. Lai, "Membrane systems engineering for post-combustion carbon capture," *Energy Procedia,* vol. 37, pp. 976–985, 2013.
87. O. San, "The digital twin revolution," *Nature Computational Science,* vol. 1, pp. 307–308, 2021.
88. J. K. Møller, G. Goranović, P. Brath, and H. Madsen, "A data-driven digital twin for water ultrafiltration," *Communications Engineering,* vol. 1, p. 23, 2022.
89. D. Jones, C. Snider, A. Nassehi, J. Yon, and B. Hicks, "Characterising the digital twin: a systematic literature review," *CIRP Journal of Manufacturing Science and Technology,* vol. 29, pp. 36–52, 2020.
90. M. Dotoli, A. Fay, M. Miśkowicz, and C. Seatzu, "Advanced control in factory automation: a survey," *International Journal of Production Research,* vol. 55, pp. 1243–1259, 2017.
91. M. G. Kapteyn, J. V. R. Pretorius, and K. E. Willcox, "A probabilistic graphical model foundation for enabling predictive digital twins at scale," *Nature Computational Science,* vol. 1, pp. 337–347, 2021.
92. M. Liu, S. Fang, H. Dong, and C. Xu, "Review of digital twin about concepts, technologies, and industrial applications," *Journal of Manufacturing Systems,* vol. 58, pp. 346–361, 2021.
93. E. Glaessgen and D. Stargel, "The digital twin paradigm for future NASA and US Air Force vehicles," in *53rd AIAA/ASME/ASCE/AHS/ASC Structures, Structural Dynamics and Materials Conference 20th AIAA/ASME/AHS Adaptive Structures Conference 14th AIAA,* Honolulu, Hawaii, p. 1818, 2012.
94. C. Ghenai, L. A. Husein, M. Al Nahlawi, A. K. Hamid, and M. Bettayeb, "Recent trends of digital twin technologies in the energy sector: a comprehensive review," *Sustainable Energy Technologies and Assessments,* vol. 54, p. 102837, 2022.
95. F. van Rooij, P. Scarf, and P. Do, "Planning the restoration of membranes in RO desalination using a digital twin," *Desalination,* vol. 519, p. 115214, 2021.
96. F. Lin, P.-C. Chen, T.-C. Huang, G.-S. Lin, and K.-L. Tung, "3D in-situ simulation and particle tracing of gas filtration process for ultrafine particles removal using a hollow fiber membrane," *Journal of Membrane Science,* vol. 632, p. 119380, 2021.
97. D. Q. F. de Menezes, M. C. C. de Sá, T. B. Fontoura, T. K. Anzai, F. C. Diehl, P. H. Thompson, et al., "Modeling of spiral wound membranes for gas separations – Part II: data reconciliation for online monitoring," *Processes,* vol. 8, p. 1035, 2020.
98. J. Fu, N. R. Ahmad, C. P. Leo, J. M. Aberilla, I. D. Cruz, B. Alamani, et al., "Techno-economic and life cycle assessment of membrane separation in post-combustion carbon capture: a review," *Gas Science and Engineering,* vol. 330, p. 205401, 2024.
99. T. Fan, S. Shen, C. Y. B. Sit, P. J. Kenis, and A. Chapman, "Environmental, economic and social trade-offs of membrane-based direct air capture technologies integrated with CO_2 conversion using life cycle assessment," *Journal of CO_2 Utilization,* vol. 91, p. 103002, 2025.

100. M. A. Budihardjo, I. F. S. Wahyuningrum, A. Chegenizadeh, A. S. Puspita, and S. Al Qadar, "Enhancing waste management in batang regency: integrating circular economy principles with life cycle assessment and material flow analysis," *Journal of Open Innovation: Technology, Market, and Complexity,* vol. 10, p. 100437, 2024.
101. Y. Yi, X. Fei, M. C. Lavagnolo, and A. Manzardo, "An integrated model of life cycle assessment and system dynamics for construction and demolition waste management and reduction in Italy," *Journal of Cleaner Production,* vol. 486, p. 144469, 2025.
102. S. S. Pattanaik, A. K. Sahoo, R. Panda, and S. Behera, "Life cycle assessment and forecasting for 30kW solar power plant using machine learning algorithms," *e-Prime-Advances in Electrical Engineering, Electronics and Energy,* vol. 7, p. 100476, 2024.
103. S. Moussavi, P. Barutha, and B. Dvorak, "Environmental life cycle assessment of a novel offshore wind energy design project: a United States based case study," *Renewable and Sustainable Energy Reviews,* vol. 185, p. 113643, 2023.
104. C. Yilmaz and S. A. Korkmaz, "Life cycle cost assessment of geothermal energy assisted hydrogen liquefaction for sustainable and renewable energy applications: case study and adaptation for Afyon geothermal power plant," *International Journal of Hydrogen Energy,* vol. 72, pp. 772–788, 2024.
105. S. Lee, W. Choi, J. H. Kim, S. Park, Y. J. Hwang, and J. Na, "Techno-economic analysis and life-cycle assessment of the electrochemical conversion process with captured CO_2 in an amine-based solvent," *Green Chemistry,* vol. 25, pp. 10398–10414, 2023.
106. S. Troy, A. Schreiber, and P. Zapp, "Life cycle assessment of membrane-based carbon capture and storage," *Clean Technologies and Environmental Policy,* vol. 18, pp. 1641–1654, 2016.
107. L. Giordano, D. Roizard, and E. Favre, "Life cycle assessment of post-combustion CO_2 capture: a comparison between membrane separation and chemical absorption processes," *International Journal of Greenhouse Gas Control,* vol. 68, pp. 146–163, 2018.
108. S. U. Hong, Y. Wang, L. S. Soh, and W. F. Yong, "Are green solvents truly green? Integrating life cycle assessment and techno-economic analysis for sustainable membrane fabrication," *Green Chemistry,* vol. 25, pp. 4501–4512, 2023.
109. R. Chauvy, D. Verdonck, L. Dubois, D. Thomas, and G. De Weireld, "Techno-economic feasibility and sustainability of an integrated carbon capture and conversion process to synthetic natural gas," *Journal of CO_2 Utilization,* vol. 47, p. 101488, 2021.
110. S. P. Nunes, P. Z. Culfaz-Emecen, G. Z. Ramon, T. Visser, G. H. Koops, W. Jin, et al., "Thinking the future of membranes: perspectives for advanced and new membrane materials and manufacturing processes," *Journal of Membrane Science,* vol. 598, p. 117761, 2020.

Section II

Growth Potential and Associated Barriers in Commercialization

4 Growth Potential of Materials for Decarbonization

Insights from the Cement Industry Value Chain

Noha A. Mostafa and Amr Elkady

4.1 INTRODUCTION

After water, concrete is the most widely used material in the world (Gagg, 2014). Concrete is the material that we interact with every day—in the structure of houses, offices, hospitals, shopping malls, supporting bridges and infrastructure, used in roads, airports, and harbors. Cement has been in use by humans throughout history; variations of the material were used up to 12,000 years ago, with the earliest archaeological discovery of consolidated whitewashed floor made from burned limestone and clay found in Palestine and Turkey (Ventolà et al., 2011). Cement and concrete are the cornerstones of the built environment. Their global demand has nearly tripled over the past 20 years, driven by the massive urban development. The global production of Portland cement increased from 3.3 billion tons in 2010 to 4.1 billion tons in 2022 (Khalil and AbouZeid, 2025), while the cement industry is responsible for approximately 7% of global CO_2 emissions (Hanifa et al., 2023). Cement basically acts as a glue that binds the ingredients of concrete together to make it a useful and multiuse building material that provides solutions around the world and will continue in the foreseen future. The cement market growth is driven by the increasing global demand for construction activities. As urbanization and population growth continue, countries invest in new infrastructure projects, driving the demand for cement in residential, commercial, and industrial constructions. In 2030, global cement production is projected to reach 4.55 billion tons, and cement consumption is anticipated to rise to 4.54 billion tons (GlobeNewswire, 2025). Figure 4.1 shows the projected global cement market size by product type from 2023 to 2033.

The cement manufacturing process starts with mining and crushing raw materials like limestone and clay to a fine powder, called raw meal, which is then gradually heated to reach a temperature at around 1450°C in the cement kiln. In this process, the chemical bonds of the raw materials are broken down and are then recombined into new compounds. in the outcome is creating a new material called clinker, in the form of rounded particles between 1 and 25 mm across. The clinker is then ground to a finer powder in the cement mills and mixed with gypsum to create the basic

DOI: 10.1201/9781003582304-6

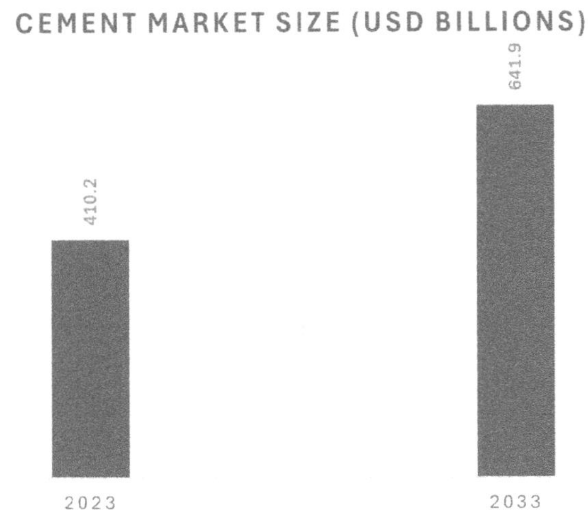

FIGURE 4.1 Global cement market forecast.

ordinary Portland cement (OPC). This fine powder cement is then mixed with water and other aggregates to form concrete. There are other types of cement used for different applications; however, OPC is the most basic and common one.

4.2 ROLE OF CEMENT INDUSTRY IN GLOBAL CARBON EMISSIONS

Cement production generates large amounts of carbon dioxide, directly generated from the calcination of limestone, the burning of fossil fuels during high-temperature calcination, electricity consumption, and emissions from the transportation and processing of raw materials. Cement plants are dispersed across the globe, with a notable concentration in Asia, particularly China, which is the world's largest cement producer. In 2023, the global carbon emissions reached 36.8 billion tons, with cement production accounting for around 7% of this total volume (Wu et al., 2024). Figure 4.2 shows the map of global cement production plants.

During the cement manufacturing process, a series of complex reactions occur, including the decarbonation of limestone and the formation of calcium silicates and aluminates, which are the main components of clinker. After leaving the kiln, the clinker is cooled to prevent the formation of undesired crystal structures. The cooled clinker is ground into a fine powder mixed with gypsum to produce the final product.

While cement demand is increasing, there is an increased pressure to meet this demand through more sustainably produced materials. Cement and concrete manufacturers can engage in the following levers to decarbonize along the value chain:

4.2.1 Reducing the Amount of Clinker in Cement

Clinker accounts for approximately 90% of the emissions released in cement production, making it a high priority for industry players to address (Shin et al., 2025). To decarbonize the cement industry, clinker can be substituted with alternative materials, such as fly ash, calcined clay, or metal slag. They can also seek to reduce the amount of clinker needed through innovations in construction, design, and material mixes and developing low-binder cements and reusable concrete modules.

4.2.2 Reducing CO_2 from Energy Used in the Production Process

To reduce energy-related emissions, alternative fuels and newer technologies, such as kiln electrification, are explored. In addition, cement and concrete producers are developing proprietary waste-recycling businesses that target the use of industrial and municipal waste in their kilns. Cement manufacturers highlighted their specific appetite for businesses that aim to capture and utilize industrial carbon emissions. Figure 4.3 shows Scope 1 and 2 emissions from producing cement.

FIGURE 4.2 Map of global cement production plants. Size of the bubbles represents the scale of CO_2 emissions. (CaptureMap endrava, 2022.)

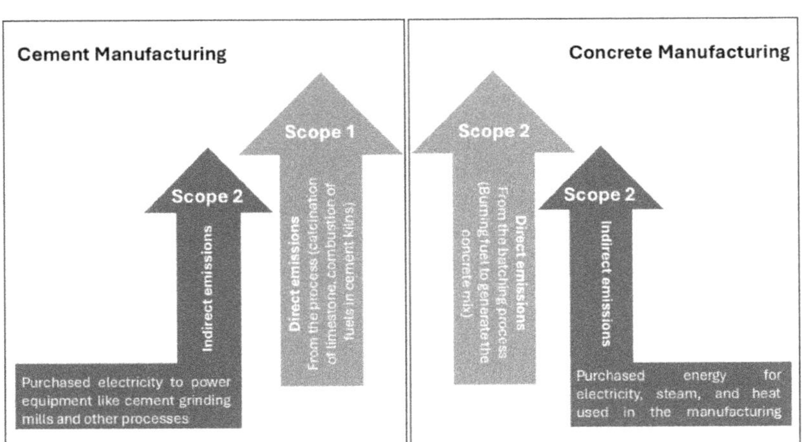

FIGURE 4.3 Scope 1 and 2 emissions from producing cement and concrete.

4.3 EMERGING SUSTAINABLE BUILDING MATERIALS

Sustainable, or green, building materials are eco-friendly products that make use of waste, reduces carbon dioxide emission across their value chains, and require less energy for production (Naqi and Jang, 2019). Low-carbon cement is an innovation that has gained momentum during the past decade to support the transition toward minimizing the environmental footprint of construction. Sustainable materials innovation can benefit the planet by positively impacting ecosystems, while supporting the well-being of people. Low-carbon building materials can be characterized by the following:

- **Sustainability**: Made from renewable, recycled, or have a minimal environmental impact throughout their life cycle and minimizing the use of natural resources.
- **Durability**: Long life and more reliable, leading to reduced need for replacements and better resources conservation over time.
- **Energy efficiency**: Contribute to reduced energy consumption by improving insulation, enhancing building performance, and supporting efficient heating and cooling systems.
- **Resource-friendly production**: This can include using renewable energy to power production processes, minimizing waste, and employing sustainable practices such as energy-efficient technologies in the manufacturing process.

4.4 DECARBONIZATION THROUGH MATERIAL SUBSTITUTION

Key materials used in low-carbon cement products include the following.

4.4.1 Fly Ash

Fly ash is a by-product of coal combustion in power plants and is considered a supplementary cementitious material (SCM) (Fořt et al., 2021). It is used in concrete mixes as a partial replacement for higher carbon-emitting Portland cement. Fly ash improves the workability and durability of concrete while reducing the overall carbon footprint and reusing an industrial by-product.

4.4.2 Slag

Slag, also known as ground granulated blast-furnace slag, is a by-product of the steel manufacturing process (Ahmad et al., 2022). It is also considered SCM and is used to replace a portion of higher carbon-emitting Portland cement in concrete mixes, enhancing durability and reducing the heat of hydration, which lowers the risk of thermal cracking.

4.4.3 Limestone

Limestone is a natural mineral used to reduce the clinker content in cement, thereby lowering CO_2 emissions per tonne of cement (Thomas et al., 2012). Finely ground limestone can also improve concrete's early strength and workability.

4.4.4 Natural Pozzolans

Natural pozzolans, such as volcanic ash, enhance the properties of green cement. These materials react with calcium hydroxide in the presence of water to form compounds that improve the strength and durability of concrete (Hamada et al., 2023).

4.4.5 Recycled Aggregates

Recycled aggregates are materials such as crushed concrete from demolition projects that are reused in new concrete mixes. This practice reduces the need for virgin aggregates and cementitious materials, and minimizes construction waste (Silva et al., 2019).

4.4.6 Silica Fume

Silica fume is a by-product of producing silicon and ferrosilicon alloys. It is a highly reactive pozzolan that, when added to concrete, increases its strength and durability while reducing permeability (Amin et al., 2022).

4.4.7 Calcined Clay

Known also as metakaolin, it is produced by heating a source of kaolinite to between 650°C and 750°C. Kaolin is both naturally occurring as well as in industrial by-products, such as some paper sludge waste and oil sands tailings (Dhar and Bishnoi, 2024). It is a pozzolanic material with smaller particle size than Portland cement. Calcined clay and limestone are already commonly used as SCM. The major innovation in LC3 is to combine the use of abundantly available low-grade kaolinite clay with a further 15% of limestone, which together have a synergetic effect and reach a similar performance as OPC.

LC3 can potentially save up to 40% of CO_2 as compared to OPC. As previously explained, most of the CO_2 comes from the clinker production process. Therefore, reducing the clinker factor and replacing it with already calcined materials is the fastest intervention to reduce CO_2. Moreover, utilization of lower-grade material for LC3 like clay waste, e.g., from ceramic or cosmetic industry, would save natural resources and enhance the circularity of waste.

4.4.8 Alternative Fuels

Known also as waste-derived fuels, alternative fuels are one way through which cement manufacturers can reduce environmental impacts and carbon emissions. The process of "Co-processing", which is the combination of simultaneous material recycling and energy recovery from waste in a thermal process, can result in replacing fossil fuels such as coal and petroleum products. Co-processing is a waste management option that is used globally, given its benefits in managing wastes. In Europe, the cement sector is substituting on average 53% of its fossil fuel with waste-derived fuels, and some cement plants already operate at about 100% substitution of fossil fuels. Globally, the figure is estimated to be around 6% showing the further potential of managing wastes in cement kilns. By using liquid waste, some plants may reach a substitution rate of up to 80% (Tihin et al., 2023; Global Cement and Concrete Association, 2024).

Co-processing reduces global CO_2 emissions and global fossil fuel consumption, as it reduces the amount of waste that is incinerated or sent to landfill (Nejad et al., 2025). Additionally, cement kilns reduce their demand for fossil fuels although the consumption of waste-derived fuel is often associated with a small but appreciable increase in kiln fuel consumption overall. Due to the high temperatures in the clinker kiln, the cement sector can substitute fossil fuels with fuels derived from waste. In countries which lack a sound waste management system, co-processing offers the opportunity to divert waste away from open burning and landfill to existing cement kilns. This reduces the demand for investment in other waste management options. However, some preconditions need to be in place. Countries must develop waste collection and waste separation systems, and regulations on co-processing; cement companies have to build up know-how and skills on co-processing, and they also need to make the necessary investments in equipment. Most countries have cement kiln production capacities, and the key advantages of using them are the reduced investment requirements and the capacity to deal with plastic and other wastes on an industrial scale from plastic hotspots or mining efforts at landfills and dumpsites. With this approach, the energy content in nonrecyclable plastic waste can be harnessed to produce clinker, while reducing the reliance on fossil fuels (Barbhuiya et al., 2024).

However, supply chains for low-carbon cement are facing various challenges. There is often a mismatch between green cement suppliers and end-consumers. Moreover, voluntary demand may be on a project-by-project basis and lack consistent volume, making long-term certainty difficult for emerging green cement players (Terán-Cuadrado et al., 2024). Taking inspiration from the power sector, there are now efforts to develop a "book and claim" system for low-carbon cement (Brander and Bjørn, 2023). This chain of custody model allows the positive environmental attributes of low-carbon cement to be purchased, even though the low-carbon cement is physically used elsewhere. Purchasers can therefore address supply chain emissions, while low-carbon cement start-ups can simplify logistics by always using its cement locally. The book and claim approach would open a much larger pool of buyers for low-carbon cement products.

4.5 USE OF CARBON CAPTURE, UTILIZATION, AND STORAGE (CCUS) TECHNOLOGIES

Another action that the cement industry is currently considering and testing to further development toward reducing its CO_2 emissions is the carbon capture and storage (CCS). Since the largest share of cement production emissions comes from process emissions and is tightly tied to the production process, emission abatement measures such as fuel switching have only a limited effect on the total CO_2 emissions from cement plants. Solutions relevant to change in the production process are studied; however, such solutions are usually complex and require more time to be realized. This is where carbon capture can help the cement industry by providing an available solution that can abate a large share of the CO_2 emissions at existing or new plants (Izumi et al., 2021).

Most cement plants can use carbon capture technology. The plant fumes typically include 15%–30% CO_2 concentration, which is higher than many other industries, making it relatively easier and cheaper to capture the CO_2 (CaptureMap, 2022). Cement plants usually have one to a few emission point which makes capture design relatively less complicated. The cement industry is already underway in testing technologies and designing carbon capture projects. The International Energy Agency carbon capture project database currently includes 27 projects for carbon capture in cement plants located mainly in Europe and North America. The announced capture capacity for these projects varies from 100,000 tons per year, up to 2.9 million tons CO_2 per year (IEA, 2024).

CCS aims to capture this CO_2 and prevent it from being released, thus mitigating climate change and reducing carbon emissions. One common approach is post-combustion capture, which involves capturing CO_2 from the flue gases generated during cement production (Galusnyak et al., 2022). This process typically employs chemical solvents or sorbents that selectively capture CO_2 from the flue gas stream. The flue gases are passed through an absorber where the CO_2 reacts with the solvent, forming a solution of CO_2-rich solvent. The solvent is then regenerated by heating, releasing the captured CO_2, which can be further purified and compressed for transport and storage. While this approach is relatively costly, it is still more suitable for making higher-value products such as steel rather than cement, by 2050, they could more than halve emissions (Mckinsey & Company, 2020). Various post-combustion carbon capture pilots are running large cement manufacturers. Other companies are testing oxyfuel combustion, a promising but expensive technology that results in high concentrations of CO_2 in flue gas, which in turn allows for near-total carbon capture (Kroumian et al., 2025). Capture costs can be reduced through using waste heat whenever available on site. This varies from plant to plant; about 70%–90% of the CO_2 can be typically captured, although some plants may choose to capture a lower share of their CO_2 (Hanifa et al., 2023).

4.6 ENERGY EFFICIENCY IN MATERIAL PROCESSING

For cement production, thermal energy accounts for 20%–25% used mainly during the burning process, while 110–120 kWh of electrical power is required to produce one ton of cement, with most of electrical power consumption is during the grinding process (Sahoo et al., 2022). Many continuous improvement efforts were made in Europe over the past ten years to decrease power consumption by investing in new equipment and retrofitting an existing one. Cement plants can have up to and more than 1000 motors in operation (including conveyors, pumps, small fans, and blowers), depending on the size of the plant and the process technology. Their substitution, or retrofitting with the application of variable speed drivers, which is becoming more and more widespread, brought a significant step forward. For further significant steps in reducing the specific energy demand, huge retrofits like replacing cement grinding with ball mills to highly efficient vertical roller mills or high-pressure grinding rolls will play a significant role (Pareek and Sankhla, 2021). The investment for these kinds of retrofits is high; therefore, they will mainly be carried out if the market situation becomes very promising or if the equipment is already obsolete and needs to be replaced.

Electrical energy demand depends on product properties. The higher the cement's strength development, the finer it typically has to be ground, requiring significantly more energy in the mills. This trend can increase the demand of electrical energy, despite the improvement in equipment efficiency. High-efficiency separators are widely used nowadays, given its impact on cement quality. It was observed that the tougher the environmental requirements, the higher. the power consumption. For example, lower values for dust emission limits require more power needed for dust separation. For the abatement of other components (e.g., NO_x or SO_2), additional units need to be installed which consume higher electricity. The use of selective catalytic reduction technology for NO_x abatement increases the electrical energy demand by 5 kWh t^{-2} (Ciobanu et al., 2022).

Looking at the future, a major reconfiguration of the production process might be needed due to the implementation of CCUS technologies. Capture technologies, such as oxyfuel as well as post-combustion technologies, will require high power consumption for oxygen production in an air separation unit, the regeneration of absorbent agents, and the separation, purification, and compression of CO_2. Therefore, carbon capture would increase power consumption by 50%–120% at the plant level (*CEMBUREAU*, the European Cement Association, 2018).

4.6.1 Renewable Power Consumption

Cement process can correct the fluctuations in the management of energy supply by renewable sources avoiding the necessity of storage capacities for electrical energy. To address this issue, producers can coordinate their electrical power-intensive processes regarding the grid operating times and duration (Mostafa et al., 2022). This leads to operating cement mills only at night, when the energy demand and prices are lower (Mossie et al., 2025). Flexible grinding approach can be used for power-intensive processes such as raw meal and cement manufacturing. The management of silo and production capacities could then be utilized similar to a battery to make optimal use of excess renewable energy. Demand decrease during periods of lack of electrical energy supply could decrease the need for power production from fossil fuels. The mills should be able to grind a maximum amount of material at times of abundant renewable energy so that the clinker production can continue without any stoppages. The main drawbacks are less efficient use of the mill and silo capacities and that the forecasting period and the peak load times can be too short for a start-up of the mill and for continuous mill operation. An on-off operation of mills would increase energy losses, wear of equipment, maintenance, and some amounts of product with uncertain product quality (Woywadt and Schmalenberger, 2022).

On-site power generation in the cement industry is another promising approach mainly in emerging markets, where some cement producers are investing in wind farms or solar plants to ensure the supply in case of grid instability and to directly use renewable energy for cement production, in particular for grinding units (Ayling et al., 2021).

Waste heat from the kiln system can be used to produce electricity. Normally waste heat (kiln exhaust gas, bypass gas, and/or cooler exhaust air) is used for the drying of raw materials or other materials like coal and petcoke, or cement constituents. It can also be partially recovered to produce electricity through a heat recovery boiler and a turbine system. Depending on the chosen process, and use of steam, ammonia, organic, and kiln, technology in total, between 8 and 22 kWh t^{-2} clinker or up to 16% of the power consumption of a cement plant can be produced by using these technologies without changes to the kiln operation (Wang et al., 2014). Main constraints for the development of these technologies are the investment costs and efficiency limited by a low-temperature level.

4.6.2 Electrification of the Production Process

Cement production requires high temperatures that can up to 1450°C; hence, using electricity for heat input is a challenge given the increasing energy bills. There are no commercially available solutions so far, mainly due to a lack of a business case for their development. While there is no intrinsic technical obstacle, a key condition for future electrification for the heat production is that the electricity is fossil-free and comes at affordable price. The possible technologies under study are as follows:

- **Plasma**: A fundamental state of matter that occurs when a gas is heated sufficient to form an ionized gas. Temperatures between 3000°C and 5000°C can be obtained. The concept of using plasma generators in a preheater, pre-calciner kiln is at present the main technology path (Parra and Romano, 2023).
- **Electrical flow heaters**: Heat is generated by running a current through a resistive element, which usually is protected by a shroud, and transferred to gas flow through high-velocity convection. Maximum gas outlet temperatures of 1100°C–1200°C are quoted (*CEMBUREAU*, the European Cement Association, 2018).
- Other methods include microwave heating, Resistive electrical heating, and Induction heating.

4.7 ROLE OF DIGITALIZATION AND DATA-DRIVEN INNOVATION

Leveraging digitalization and artificial intelligence (AI) to accelerate sustainability can provide cement producers with new tools to drive decarbonization, including AI-based optimizations, Internet of Things, and data analytics. Some of the promising digital solutions that can be adopted in the cement industry are as follows.

4.7.1 Predictive Maintenance

Poor maintenance often leads to a 5%–20% downgrade in a plant's overall productive capacity (Cement Manufacturers Association, 2021). Predictive maintenance allows plant operators to monitor each installation's integrity and assess the probability of breakdown, which enables timely addressal of maintenance issues. Besides preventing ling downtime, it can achieve higher operational efficiency, less maintenance cost, and less wear and tear of machinery. In predictive maintenance, the operator halts production for limited time each month to replace equipment based on its health as opposed to routine periodic replacements. Predictive maintenance can be particularly useful for gearboxes and kiln-related installations.

4.7.2 Digital Twin

A digital twin allows cement companies to mirror their entire production process or a portion thereof through a digital model, referred to as twin of the plant's physical assets, processes, and systems. The twin generates scenarios that vary across process steps and variables. The model proposes optimal equipment settings for best performance. The solution leverages use of AI and machine learning (ML) to optimize the manufacturing process. Such models are dynamic and can learn and improve with each experience (Tong et al., 2023). As the model evolves, users get data-driven predictions without need for frequent reprogramming. In 2021, the global digital twin industry had a value of 6.5 billion USD, with a projected growth to 124.7 billion USD by 2030 (Yakovenko and Shaptala, 2024).

4.7.3 Predictive Quality Analytics (PQA)

PQA enable cement companies to predict cement quality with higher accuracy. The model helps manufacturers to choose among a multitude of alternative approaches, such as making use of high-grade limestone, additives, or extensive milling for fineness, among others, to produce a desired quality product. A range of alternative tools can be used based on the business model and the unique competitive advantages. PQA models use ML algorithms to correlate the quality of each production batch with the relevant production parameters (Ismail et al., 2022; Barrionuevo et al., 2025). The use of AI enables the system to learn over a period and to associate all the process input parameters with the quality outcome. As such, the producers can predict the physical and chemical properties of the cement and address any shortcomings (Ali et al., 2022).

4.7.4 Alternative Fuel Optimizer

The use of alternative fuels is identified as a potent decarbonization lever and can significantly help reduce the expenditure on energy in cement production. Alternative fuel optimizer enables cement manufacturers to manage the inherent variability of alternative fuels, specifically its physical and chemical properties such as calorific value and compatibility of fuels (Luo et al., 2024).

4.7.5 Integrated Control Tower

An integrated control tower (ICT) allows operators to remotely run their cement plants from a central office, thereby reducing operational and maintenance costs, optimizing workforce, and improving health and safety performance. ICT can perform condition monitoring for equipment, predict failures, support maintenance planning and optimal selection of settings and inputs, usually through scenario analysis software. A control tower can enable a company to realize a 5%–10% improvement in on-time delivery performance and/or a 10%–20% reduction in distribution costs (Cement Manufacturers Association, 2021; Gerrits et al., 2022).

4.7.6 Digitalization of Cement Value Chain

Digitizing upstream and downstream supply chain helps in improving visibility on raw materials, fuels, and finished goods, deliver quality customer service, and optimize cost (Walther, 2018). Tagging trucks with Global Positioning System helps in encouraging responsible behavior among drivers, minimizing deviation from set paths, thereby reducing turnaround time and safety incidents. Digitalization of consumer purchase cycle eases the management of construction projects for consumers. This cycle includes order booking, sharing acknowledgments, tracking, monitoring of delivery, and payment receipts.

4.7.7 Circular Economy and Recycled Materials

Incorporating circular economy in construction involves selecting materials that are not only durable and flexible but also designed for reusability. This approach conserves valuable resources and initiates innovative avenues in the building design. It also emphasizes that materials used today will still be valuable in the future to help support a sustainable construction industry (Özçelikci et al., 2023). Strategies such as reducing energy consumption, optimizing designs, employing low-carbon materials, and embracing digital technologies contribute to this goal. Nonetheless, it is imperative to adopt circular economy principles, wherein buildings are conceived to be assembled, utilized, disassembled, and repurposed as necessary. This approach not only aligns with the net-zero agenda but also prioritizes waste reduction in building design.

Circular construction entails the creation, utilization, and repurposing of buildings, construction elements (e.g., materials, spaces, and infrastructure), all while minimizing the depletion of natural resources, environmental pollution, and negative impacts on ecosystems. This process starts way before the cement is produced; it starts from the design of a structure or a building. Buildings should be designed for longevity, adaptability, and disassembly, integrating that it should be built efficiently and with recycled, renewable, and nonpolluting materials. Using circular building materials achieves substantial economic and environmental benefits. Economically, cost savings can be achieved through reduced costs of material purchases and waste disposal. Environmentally, using circular materials minimizes the depletion of natural resources and reduces greenhouse gas emissions.

4.7.8 Supply Chain/Value Chain Collaboration

Building materials account for half of the solid waste generated every year worldwide with more than 3.57 billion tons of construction and demolition (C&D) waste produced annually (Patil et al., 2024). One way to reduce the amount of waste and introduce circularity into the value chain is by recycling C&D waste to produce concrete. This has other promising applications especially in regions affected by war such as Gaza that was heavily destructed during the war on Palestine and Ukraine that has various regions with damaged buildings (Emmanuel et al., 2025). Other solutions include introducing new materials that can be easily deconstructed and reused in other buildings once the original building is torn down and expanding the use of carbon calculators such as EC3 to give visibility to the embodied carbon emissions. This would allow building owners, green building certification programs, and policymakers to assess supply chain data to establish requirements and set embodied carbon limits at the project stage.

4.8 POLICY AND REGULATION IMPACT

Many of the decarbonization solutions discussed earlier in this chapter require major investments in infrastructure, alternative energy sources, and new technologies, including for CCUS. Cement producers have been unenthusiastic to widely adopt those without assurance or financial motivation of their economic and technical feasibility. Despite the advancements toward more sustainable practices, CO_2 pricing policies are still insufficient to pressure manufacturers to upgrade their facilities (Cleary and Willcott, 2024). Governments can play a substantial role by providing subsidies, policy support, and regulation. A cement-specific approach to government intervention can be taken, with one example being China's National Development and Reform Commission plan to cap clinker production, improve thermal efficiency, and increase the use of alternative fuels by implementing incentives and giving financial support to cement plants (Li et al., 2025). The European Union instead includes cement decarbonization in regulations covering several industries such as the Emissions Trading System and the Carbon Border Adjustment Mechanism (Erdogdu, 2025). Billions have been allocated to cement-sector carbon capture projects in the world with many success stories (Guo et al., 2024). Many innovative carbon capture approaches were demonstrated within the cement sector, including oxyfuel combustion and hybrid pressure-swing adsorption with cryogenic distillation.

Green public procurement is powerful tool for governments that usually buy massive amounts of cement for infrastructure and national projects. At COP28 in December 2023, the governments of Canada, Germany, the United Kingdom, and the United States pledged to adopt timebound commitments to procure low-emission steel, cement, and concrete (IDTechEx, 2025). Standards, such as green building codes, supports and motivates the adoption of sustainable cement and concrete practices (Barbhuiya et al., 2024).

4.9 CONCLUSIONS

Businesses operating in the built industry are increasingly becoming active in the low-carbon cement space, as seeking deals with cement players developing low-carbon products helps generate positive public and market image. Key trends in cement industry include innovations in materials and operations to reduce carbon footprints, government policies, rising demand for green buildings, and expansion of green cement technologies.

The market of lower-carbon cement is expected to grow over the coming years with more countries shifting toward sustainability which still needs more research. Another future research direction is using low-carbon cement as a solution to decarbonize data centers.

REFERENCES

Ahmad, J., Kontoleon, K. J., Majdi, A., Naqash, M. T., Deifalla, A. F., Ben Kahla, N., Isleem, H. F., and Qaidi, S. M. A. A comprehensive review on the Ground Granulated Blast Furnace Slag (GGBS) in concrete production. *Sustainability*, 14, 8783, 2022. https://doi.org/10.3390/su14148783

Ali, A. M., Tabares, J. D., and McGinley, M. W. A machine learning approach for clinker quality prediction and nonlinear model predictive control design for a rotary cement kiln. *Journal of Advanced Manufacturing and Processing*, 4(4), e10137, 2022. https://doi.org/10.1002/amp2.10137

Amin, M., Zeyad, A. M., Tayeh, B. A., and Agwa, I. S. Effect of ferrosilicon and silica fume on mechanical, durability, and microstructure characteristics of ultra high-performance concrete. *Construction and Building Materials*, 320, 126233, 2022. https://doi.org/10.1016/j.conbuildmat.2021.126233

Ayling, G., Domun, A., and de Villa, G. Pathways to alternative power and decarbonization technologies in cement manufacturing. *2021 IEEE-IAS/PCA Cement Industry Conference (IAS/PCA)*, Orlando, FL, USA, 1–17, 2021. https://doi.org/10.1109/IAS/PCA50164.2021.10143787

Barbhuiya, S., Das, B. B., and Adak, D. Roadmap to a net-zero carbon cement sector: strategies, innovations and policy imperatives. *Journal of Environmental Management*, 359, 121052, 2024. https://doi.org/10.1016/j.jenvman.2024.121052

Barrionuevo, R., Vallejo-Huanga, D., Morillo, P., and Mostafa, N. A. Predictive quality analytics for the viscosity of water-based architectural paint manufacturing by using improved supervised machine learning and maximum dissimilarity algorithm. *Journal of Intelligent Manufacturing*, 2024. https://doi.org/10.1007/s10845-025-02594-5

Brander, M., and Bjørn, A. Principles for accurate GHG inventories and options for market-based accounting. *The International Journal of Life Cycle Assessment*, 28, 1248–1260, 2023. https://doi.org/10.1007/s11367-023-02203-8

CaptureMap endrava. *The Cement Industry Needs Carbon Capture for Reducing Emissions*, 2022. https://www.capturemap.no/carbon-capture-cement-industry/

CEMBUREAU, the European Cement Association, 2018. https://lowcarboneconomy.cembureau.eu/5-parallel-routes/energy-efficiency/electrical-energy-efficiency/

Cement Manufacturers Association, Digitalisation in Cement Industry, 2021. https://www.cmaindia.org/digitalisation-in-cement-industry

Ciobanu, C., Tudor, P., Istrate, I. -A., and Voicu, G. Assessment of environmental pollution in cement plant areas in Romania by co-processing waste in clinker kilns. *Energies*, 15, 2656, 2022. https://doi.org/10.3390/en15072656

Cleary, S., and Willcott, N. Carbon pricing: necessary but not sufficient. *Finance Research Letters*, 68, 106017, 2024. https://doi.org/10.1016/j.frl.2024.106017

Dhar, M., and Bishnoi, S. Influence of calcination temperature on the physical and chemical characteristics of kaolinitic clays for use as supplementary cementitious materials. *Cement and Concrete Research*, 178, 107464, 2024. https://doi.org/10.1016/j.cemconres.2024.107464

Emmanuel, O., Białko, M., and Nikolaiev, V. The construction industry in a war-ravaged region: examination of challenging factors. *Applied Sciences*, 15, 2900, 2024. https://doi.org/10.3390/app15062900

Erdogdu, E. The carbon border adjustment mechanism: opportunities and challenges for non-EU countries. *WIREs Energy and Environment*, 14, e70000, 2024. https://doi.org/10.1002/wene.70000

Fořt, J., Šál, J., Ševčík, R., Doleželová, M., Keppert, M., Jerman, M., Záleská, M., Stehel, V., and Černý, R. Biomass fly ash as an alternative to coal fly ash in blended cements: functional aspects. *Construction and Building Materials*, 271, 121544, 2021. https://doi.org/10.1016/j.conbuildmat.2020.121544

Gagg, C. R. Cement and concrete as an engineering material: an historic appraisal and case study analysis. *Engineering Failure Analysis*, 40, 114–140, 2014. https://doi.org/10.1016/j.engfailanal.2014.02.004

Galusnyak, S. C., Petrescu, L., and Cormos, C.-C. Environmental impact assessment of post-combustion CO_2 capture technologies applied to cement production plants. *Journal of Environmental Management*, 320, 115908, 2022. https://doi.org/10.1016/j.jenvman.2022.115908

Gerrits, B., Topan, E., and van der Heijden, M. C. Operational planning in service control towers – heuristics and case study. *European Journal of Operational Research*, 302(3), 983–998, 2022. https://doi.org/10.1016/j.ejor.2022.01.025

Global Cement and Concrete Association. *GCCA Policy Document on Co-Processing*, 2024.

Globenewswire. Cement Market Global Market Report 2025**,** 2024. https://www.globenewswire.com/news-release/2025/04/09/3058654/28124/en/Cement-Market-Global-Market-Report-2025-Asia-Pacific-Dominates-Cement-Production-Amid-Urbanization-Surge.html

Guo, Y., Luo, L., Liu, T., Hao, L., Li, Y., Liu, P., and Zhu, Y. A review of low-carbon technologies and projects for the global cement industry. *Journal of Environmental Sciences*, 136, 682–697, 2024. https://doi.org/10.1016/j.jes.2023.01.021

Hamada, H. M., Abed, F., Beddu, S., Humada, A. M., and Majdi, A. Effect of volcanic ash and natural pozzolana on mechanical properties of sustainable cement concrete: a comprehensive review. *Case Studies in Construction Materials*, 19, e02425, 2023. https://doi.org/10.1016/j.cscm.2023.e02425

Hanifa, M., Agarwal, R., Sharma, U., Thapliyal, P. C., and Singh, L. P. A review on CO_2 capture and sequestration in the construction industry: emerging approaches and commercialised technologies. *Journal of CO_2 Utilization*, 67, 102292, 2023. https://doi.org/10.1016/j.jcou.2022.102292

IDTechEx. *Demand for Low-Carbon Cement is on the Rise*, 2024. https://www.idtechex.com/en/research-article/demand-for-low-carbon-cement-is-on-the-rise/32355

IEA. *Direct Air Capture*, 2024. https://www.iea.org/energy-system/carbon-capture-utilisation-and-storage/direct-air-capture

Ismail, M., Mostafa, N., and Elassal A. Quality monitoring in multistage manufacturing systems by using machine learning techniques. *Journal of Intelligent Manufacturing*, 33(8), 2471–2486, 2022. https://doi.org/10.1007/s10845-021-01792-1

Izumi, Y., Iizuka, A., and Ho, H.-J. Calculation of greenhouse gas emissions for a carbon recycling system using mineral carbon capture and utilization technology in the cement industry. *Journal of Cleaner Production*, 312, 127618, 2021. https://doi.org/10.1016/j.jclepro.2021.127618

Khalil, E., and AbouZeid, M. Framework for cement plants assessment through cement production improvement measures for reduction of CO_2 emissions towards net zero emissions. *Construction Materials*, 5, 20, 2024. https://doi.org/10.3390/constrmater5020020

Kroumian, C., Maier, J., Peloriadi, K., Scheffknecht, G., and Grammelis, P. Evaluation of 100% alternative fuel combustion under oxyfuel conditions in a pilot-scale burner for application in retrofit oxyfuel cement kiln. *Fuel*, 381(D), 133697, 2024. https://doi.org/10.1016/j.fuel.2024.133697

Li, J., Li, Y., Sun, M. et al. Studying the cement industry's policy incentives and the optimization path for carbon neutrality technology based on an integrated model. *Environment, Development and Sustainability*, 2024. https://doi.org/10.1007/s10668-025-05991-1

Luo, Z., Song, H., Huang, Y., and Jin, B. Recent advances on the uses of biomass alternative fuels in cement manufacturing process: a review. *Energy & Fuels*, 38(9), 7454–7479, 2024. https://doi.org/10.1021/acs.energyfuels.3c04535

Mckinsey & Company. *Laying the Foundation for Zero-Carbon Cement*, 2020. https://www.mckinsey.com/industries/chemicals/our-insights/laying-the-foundation-for-zero-carbon-cement

Mostafa, N., Grida, M., Park, J., and Ramadan, H. A sustainable user-centered application for residential energy consumption saving. *Sustainable Energy Technologies and Assessments*, 53(D), 102754, 2022. https://doi.org/10.1016/j.seta.2022.102754

Mossie, A. T., Khatiwada, D., Palm, B., and Bekele, G. Energy demand flexibility potential in cement industries: how does it contribute to energy supply security and environmental sustainability? *Applied Energy*, 377(C), 124608, 2024. https://doi.org/10.1016/j.apenergy.2024.124608

Naqi, A., and Jang, J. G. recent progress in green cement technology utilizing low-carbon emission fuels and raw materials: a review. *Sustainability*, 11, 537, 2019. https://doi.org/10.3390/su11020537

Nejad, B. M., Enferadi, S., and Andrew, R. A comprehensive analysis of process-related CO_2 emissions from Iran's cement industry. *Cleaner Environmental Systems*, 16, 100251, 2025. https://doi.org/10.1016/j.cesys.2024.100251

Özçelikci, E., Oskay, A., Bayer, İ. R., and Şahmaran, M. Eco-hybrid cement-based building insulation materials as a circular economy solution to construction and demolition waste. *Cement and Concrete Composites*, 141, 105149, 2023. https://doi.org/10.1016/j.cemconcomp.2023.105149

Pareek, P., and Sankhla, V. S. Review on vertical roller mill in cement industry & its performance parameters. *Materials Today: Proceedings*, 44(6), 4621–4627, 2021. https://doi.org/10.1016/j.matpr.2020.10.916

Parra, S. Q., and Romano, M. C. Decarbonization of cement production by electrification. *Journal of Cleaner Production*, 425, 138913, 2023. https://doi.org/10.1016/j.jclepro.2023.138913

Patil, Y. R., Dakwale, V. A., and Ralegaonkar, R. V. Recycling construction and demolition waste in the sector of construction. *Advances in Civil Engineering*, 6234010, 2024. https://doi.org/10.1155/2024/6234010

Sahoo, N., Kumar, A., and Samsher. Review on energy conservation and emission reduction approaches for cement industry. *Environmental Development*, 44, 100767, 2022. https://doi.org/10.1016/j.envdev.2022.100767

Shin, S., Kim, G., Kim, J., Kim, H., and Park, S. Assessment of the applicability of waste concrete fine powder as a raw material for cement clinker. *Recycling*, 10, 17, 2024. https://doi.org/10.3390/recycling10010017

Silva, R.V., de Brito, J., and Dhir, R. K. Use of recycled aggregates arising from construction and demolition waste in new construction applications. *Journal of Cleaner Production*, 236, 117629, 2019. https://doi.org/10.1016/j.jclepro.2019.117629

Terán-Cuadrado, G., Tahir, F., Nurdiawati, A., Almarshoud, M. A., and Al-Ghamdi, S. G. Current and potential materials for the low-carbon cement production: life cycle assessment perspective. *Journal of Building Engineering*, 96, 110528, 2024. https://doi.org/10.1016/j.jobe.2024.110528

Tihin, G. L., Mo, K. H., Onn, C. C., Ong, H. C., Taufiq-Yap, Y. H., and Lee, H. V. Overview of municipal solid wastes-derived refuse-derived fuels for cement co-processing. *Alexandria Engineering Journal*, 84, 153–174, 2023. https://doi.org/10.1016/j.aej.2023.10.043

Titan Cement. *Use of Alternative Materials and Fuels*. https://www.titan.gr/en/sustainability/environment/circular-economy/alternative-raw-materials-and-fuels

Thomas, M., Barcelo, L., Blair, B., Cail, K., Delagrave, A., and Kazanis, K. Lowering the carbon footprint of concrete by reducing clinker content of cement. *Transportation Research Record*, 2290(1), 99–104, 2012. https://doi.org/10.3141/2290-13

Tong, R., Sui, T., Feng, L., and Lin, L. The digitization work of cement plant in China. *Cement and Concrete Research*, 173, 107266, 2023. https://doi.org/10.1016/j.cemconres.2023.107266

Ventolà, L., Vendrell, M., Giraldez, P., and Merino, L. Traditional organic additives improve lime mortars: new old materials for restoration and building natural stone fabrics. *Construction and Building Materials*, 25, 3313–3318, 2011. https://doi.org/10.1016/j.conbuildmat.2011.03.020

Walther, T. Digital transformation of the global cement industry. *2018 IEEE-IAS/PCA Cement Industry Conference (IAS/PCA)*, Nashville, TN, USA, 1–8, 2018. https://doi.org/10.1109/CITCON.2018.8373101

Wang, Y., Höller, S., Viebahn, P., and Hao, Z. Integrated assessment of CO_2 reduction technologies in China's cement industry. *International Journal of Greenhouse Gas Control*, 20, 27–36, 2014. https://doi.org/10.1016/j.ijggc.2013.10.004

Woywadt, C., and Schmalenberger, J. Flexibility with a modular grinding plant – case study. *2022 IEEE-IAS/PCA Cement Industry Conference (IAS/PCA), Las Vegas*, NV, USA, 1–6, 2022. https://doi.org/10.1109/IAS/PCA51038.2022.9893479

Wu, S., Shao, Z., Andrew, R.M., Bing, L., Wang, J., Niu, L., Liu, Z., and Xi, F. Global CO_2 uptake by cement materials accounts 1930–2023. *Scientific Data*, 11, 1409, 2024. https://doi.org/10.1038/s41597-024-04234-8

Yakovenko, Y., and Shaptala, R. Study of digital twins as the driving force of digital transformation and achieving the goals of sustainable development. *Technology Audit and Production Reserves*, 2(4), 11–20, 2024. https://doi.org/10.15587/2706-5448.2024.301423

5 Innovative Materials for Net-Zero Applications

Siqi Wang, Ziqi Shen, and Yongliang Yan

5.1 ADVANCED MATERIALS FOR CARBON CAPTURE

Carbon Capture, Utilisation, and Storage is a set of technologies to capture CO_2 from the large CO_2 emitters and to either reuse or store it. The first generation of carbon capture technologies (pre-, oxy-fuel, and post-combustion) are mature technologies, but the high cost and energy penalty limit its deployment. Developing high-performance carbon capture materials can potentially reduce the costs and improve its energy efficiency. This section reviews the innovative materials research in metal–organic frameworks (MOFs), covalent organic frameworks (COFs), and hybrid materials for CO_2 capture and their challenges and opportunities in the application of large-scale CO_2 capture.

5.1.1 Metal–Organic Frameworks (MOFs)

MOFs (also known as porous coordination polymers, PCPs) are crystalline materials composed of metal ions or clusters coordinated to organic ligands, forming one-, two-, or three-dimensional structures. The versatility in the choice of metal nodes and organic linkers allows for the design of MOFs with tailored pore sizes, surface areas, and functionalities [1]. One of the most remarkable features of MOFs is their exceptionally high surface area, often 1000–10,000 m^2g^{-1}, high porosity, tuneable structures, and flexible tailorability, which is significantly higher than that of traditional porous materials like zeolites and activated carbons [2]. This high surface area, combined with the ability to functionalise the pores, makes MOFs ideal for CO_2 separation applications. Figure 5.1 summarises recent achievements in CO_2 capture by MOFs. Many MOFs (e.g., Mg-MOF-74, Mn-DABDC, and MOF-177) have been reported to show a superior CO_2 uptake capacity.

The open Mg^{2+} centres are essential for the high affinity for CO_2. Yang et al. [3] reported Mg-MOF-74(S) prepared in 1 h by sonochemical method has the Brunauer–Emmett–Teller (BET) surface area of 1640 m^2g^{-1} with the high adsorption capacity (350 mg g^{-1} at 298 K). A series of mmen-M_2 (dobpdc) (mmen = N,N′-dimethylethyl enediamine, M = Mg, Mn, Fe, Co, Zn, dobpdc4− = 4,4′-dioxidobiphenyl-3,3′-dicar boxylate) have been studied as the designated ‘phase-change’ adsorbents, which has very large CO_2 separation capacities with small temperature swings [4]. Asghar et al. [5] reported a manganese-based MOF (Mn-DABDC (DABDC = diaminobenzenedicarboxylate, or 2,5-diaminoterephthalate)) via the electrochemical synthesis that can achieve high carbon dioxide uptake (92.4 wt.% at 15 bar and 273 K). UiO-66-NH_2 synthesised by the sonochemical method can achieve 3.2 mmol g^{-2} of CO_2 adsorption

DOI: 10.1201/9781003582304-7

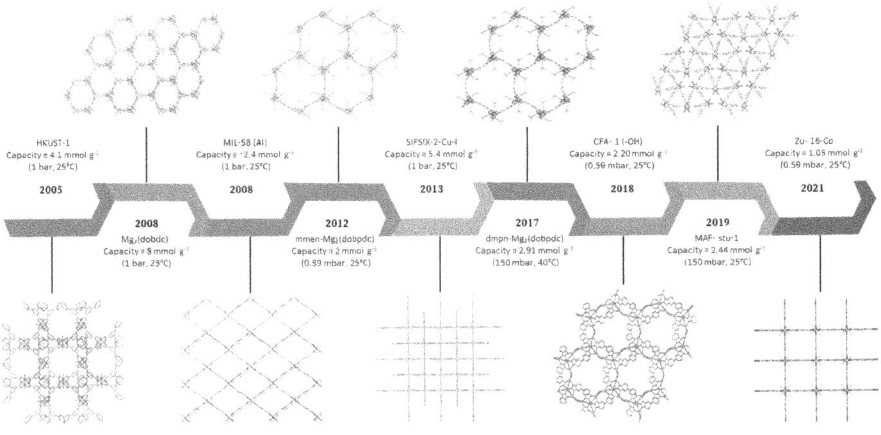

FIGURE 5.1 Research highlights in MOF adsorbents for carbon capture [7]

capacity, when compared to conventional solvothermal techniques (2.3 mmol g^{-2}) at 25°C and 1 bar, and about 0.6 mmol g^{-2} capacity decrease for sonochemical UiO-66-NH_2 throughout eight consecutive cycles [6].

However, MOFs face stability challenges, particularly under humid conditions. Many MOFs are susceptible to hydrolysis, which can lead to the collapse of their framework structure [8]. To address this issue, researchers have developed strategies such as the introduction of hydrophobic functional groups or the use of more stable metal–ligand combinations. Over the past decade, hundreds of water-stable MOFs such as MILs, UiOs, PCNs, BUTs, ZIFs, and MAFs have been developed [9]. Among them, MIL-120 is a promising water-stable and cost-effective MOF that can selectively capture the CO_2 wet flue gas [10]. Fracaroli et al. [11] reported that IRMOF-74-III-CH_2NH_2 exhibits exceptional efficiency in CO_2 uptake, achieving 3.2 mmol of CO_2 per gram at 800 Torr. More notably, it effectively removes CO_2 from humid nitrogen gas streams, demonstrating a breakthrough time of 610 ± 10 s g^{-1} while maintaining the integrity of its structure.

Recently, artificial intelligence (AI) and machine learning (ML) have been widely applied into MOFs to accelerate the design and discovery of high-performing adsorbents and membranes for CO_2 adsorption and separation [12]. Yao et al. [13] developed a deep learning algorithm for the automated design of MOF materials starting from the computation-ready, experimental MOF database and generating new proposed structures with improved properties for CO_2 separations. Park et al. [14] proposed a generative AI model – GHP-MOF assemble, which can generate novel linkers, assembled with one of three preselected metal nodes (Cu paddlewheel, Zn paddlewheel, Zn tetramer) into MOFs with high CO_2 adsorption capacity.

5.1.2 Covalent Organic Frameworks (COFs)

COFs are a class of porous materials composed entirely of light elements (such as C, H, O, N, or B atoms) connected by strong covalent bonds. Unlike MOFs, which rely on coordination bonds between metal ions and organic linkers, COFs are constructed through reversible covalent reactions such as boronate ester formation, imine condensation, and Schiff base reactions [15]. This reversible nature allows for the formation of highly ordered crystalline structures with well-defined pore sizes and geometries. One of the key advantages of COFs is their high thermal and chemical stability, which is attributed to the strong covalent bonds that hold the framework together. This stability, combined with their low density and high surface area, makes COFs attractive for applications in CO_2 separation. So far, boron, triazine, imine, and hybrid boron/imine-based COFs have been intensively investigated, and theoretical simulations have been an effective tool to simulate the CO_2 capture performance of COFs and guide its design and synthesis.

Furukawa and Yaghi [16] studied seven porous COFs (COF-1, COF-5, COF-6, COF-8, COF-10, COF-102, and COF-103) for hydrogen, methane, and carbon dioxide storage, and the COF-102 has the best CO_2 uptake capacity (1180 mg g^{-2} at 298 K and 35 bar). Zhao et al. [17] developed a perfluorinated covalent triazine-based framework (FCTF-1) for selective and water-tolerant CO_2 capture and found that the incorporation of fluorine groups could improve the COF's CO_2 adsorption and separation capacities. Nitrogen-rich COFs are promising candidates for CO_2 capture due to the increased interaction affinity of nitrogen atoms with CO_2. Kaleeswaran et al. [18] studied imine- and β-ketoenamine-based COFs, which are nitrogen-rich organic porous materials, and found that *i*PrTAPB-TFP has the highest BET surface area of 756 $m^2 g^{-1}$ and CO_2 uptake capacity of 105 mg g^{-2} at 273 K and 1 atm. Zhou et al. [19] investigated COF-999 (COF with olefin linkages) for CO_2 capture from open air, and its CO_2 capture capacity is 0.96 mmol g^{-2} at dry condition and 2.05 mmol g^{-2} under 50% relative humidity with reasonable cycling stability and low regeneration temperature (60°C). Ge et al. [20] studied the impact of water adsorption on a set of chemically stable COFs using grand canonical Monte Carlo (GCMC) simulations and found that NPN-COFs have good CO_2 capture performance with the water tolerance due to the small pores with hydrophobic surface. Qu et al. [21] also applied the GCMC method to simulate the CO_2 adsorption capacities of X-TpBD-COFs (X=$-NO_2$, $-CN$, $-Br$, $-CH_2NH_2$, $-CH_3$, $-OCH_3$) and successfully synthesise NO_2-TpBD-COF and OCH_3-TpBD-CO, which is consist to the simulations (37.74 and 17.91 $cm^3 g^{-1}$). A new framework that combines the ML and GCMC has been developed to screen 268,687 COFs for their ability to separate CO_2 from the flue gas of power plants by pressure swing adsorption process and identifies imide, (keto)enamine, and (acyl) hydrazone COFs as particularly promising for carbon capture [22]. Moreover, the highest-performing materials are 3D COFs with strong CO_2 adsorption sites located between aromatic rings on opposite sides of pores measuring 1.0 nm in diameter. In 2D COFs, the framework chemistry plays a crucial role, with features such as imide linkages and fluoro groups having a significant impact. Deeg et al. [23] developed a genetic algorithm to screen a database of more than 69,000 hypothetical COFs for carbon capture using parasitic energy as a metric and found that more than 70 COFs are suitable for CO_2 capture with the comparable CO_2 performance of Mg-MOF-74.

However, COFs face challenges related to their synthesis and scalability. The formation of highly ordered crystalline COFs often requires precise control over reaction conditions, which can be difficult to achieve on a large scale. Additionally, the reversible nature of the covalent bonds used in COF synthesis can lead to the formation of defects or incomplete frameworks [24].

5.1.3 Hybrid Materials

One single type of carbon capture material may not meet the full criteria of the selection of suitable materials for CO_2 capture process. Thus, hybrid materials can be designed to exhibit synergistic effects, where the individual components work together to achieve properties that are not attainable with a single material.

Permeability and stability issues could be overcome by incorporating filler particles such as MOFs, mesoporous silica, zeolites, and carbon-based materials into polymeric matrices [25]. Rodenas et al. [26] proposed a novel synthesis strategy that incorporates MOF nanosheets into polymer matrices, which shows a significant CO_2 separation performance from CO_2/CH_4 gas mixtures. Qian et al. made a hybrid adsorbent by incorporating HKUST-1 into hierarchical porous carbon monoliths (HCM) using a repeated impregnation and crystallisation approach and found that it enhances the CO_2 adsorption performance of the HCM. Yang et al. [27] proposed a core–shell LiX@ZIF-8 adsorbent with the combination of the high CO_2 adsorption capacity of zeolite LiX and the hydrophobic nature of ZOF-8. Seoane et al. [28] provided a critical review on MOF-based mixed matrix membranes (MMMs) for highly efficient CO_2 separation. Cheng et al. [29] studied a novel MMMs containing MOF@COF hybrid fillers for efficient CO_2 separation and found that the COF coating layer can hugely enhance the polymer–filler interfacial compatibility in MOF@COF-based MMMs. Gebremariam et al. [25] developed a novel MOF@MOF core–shell hybrid adsorbent that can improve the stability and CO_2 capture performance in moist conditions. Feitosa et al. [30] evaluated a series of organic–inorganic hybrids using different solvents for 3-aminopropyltriethoxysilane (APTS) grafting onto SBA-15, and their CO_2 adsorption capacity varied from 25 to 516 μmol g^{-2}.

5.1.4 Summary

The research in MOFs, COFs, and hybrid materials for CO_2 separation has made significant progress over the past two decades. Those materials have the remarkable properties such as high surface area and porosity, chemical and physical versatility, and chance for structural modification. Despite the remarkable progress, several challenges remain to be addressed. For MOFs, issues related to stability, particularly under humid conditions, need to be overcome to enable their widespread use in practical applications. For COFs, the challenges related to synthesis and scalability need to be addressed to facilitate their large-scale production. For the hybrid materials, the design and optimisation of these materials for CO_2 separation require a deeper understanding of the interactions between the individual components.

Future research in this field is likely to focus on the development of new materials with enhanced properties and testing the performance of suitable candidates under the industrial conditions. The use of computational methods, such as density functional theory (DFT) and molecular dynamics simulations with the integration of AI, is expected to play a key role in the design and optimisation of new MOFs, COFs, and hybrid materials. Moreover, the development of new characterisation techniques, such as in situ spectroscopy and microscopy, will provide deeper insights into the structure–property relationships of these materials.

5.2 ADVANCED MATERIALS FOR LOW-CARBON HYDROGEN PRODUCTION AND STORAGE

Hydrogen is a highly versatile feedstock extensively utilised in oil refining, ammonia synthesis, and the production of various chemicals. The growing interest in low-carbon hydrogen as an energy carrier stems from its potential to replace carbon-intensive fossil fuels, providing a pathway for decarbonising key sectors, including transport, residential heating, and power generation. In the current stage, most hydrogen came from fossil fuels, which contain a massive amount of carbon. Water electrolysis (green hydrogen production) and carbon capture and storage (CCS) coupled with conventional hydrocarbon reforming (blue hydrogen production) were recognised as the two main technologies for low-carbon hydrogen production. This section summarises the development of advanced materials used in these available technologies, as well as options for hydrogen storage.

5.2.1 Advanced Electrocatalysts for Green Hydrogen

5.2.1.1 Proton Exchange Membrane (PEM) Electrolysis

Proton exchange membrane (PEM) electrolysis is one of the most important technologies for green hydrogen production. In PEM electrolysis, a solid polymer electrolyte is used to conduct protons from the anode to the cathode, where hydrogen is produced [31]. The electrocatalysts used in PEM electrolysers can be divided into two categories, catalysts for the hydrogen and oxygen evolution reactions (HER and OER, respectively). A diagram illustrating the promotion effect of electrocatalysts on water electrolysis is shown in Figure 5.2. As shown in the figure, the activation energy requirement for splitting water into H_2 and O_2 is relatively high without an electrocatalyst. With the presence of electrocatalysts on the anode and cathode, the energy barrier for splitting water can be significantly reduced to improve the overall hydrogen production efficiency [32,33].

Pt is the most commonly used cathode electrocatalyst due to its high HER activity. In PEM electrolysers, Pt is often employed in the form of nanoparticles to the maximum number of available active sites. Due to the high cost of Pt, research has also been carried out to evaluate the performance of Pt-based alloys. Pt-Ni [34,35], Pt-Co [36,37], and Pt-free catalysts [38,39] were found to be cost-effective alternatives to pure Pt electrocatalysts with a similar level of activity and improve durability. On the other hand, OER electrocatalysts on the anode's side are primarily centred around Ir- and Ru-based materials. While IrO_x catalysts are widely used due to its stability, RuO_x have been shown to have a higher intrinsic activity [31]. Ir and Ru were also found to have a synergistic effect, as $Ir_{0.7}Ru_{0.3}O_x$ nanocrystals exhibited enhanced activity compared to their monometallic counterparts, achieving a current density of 3.2 A cm^{-2} at 1.85 V and 90°C [40]. To further enhance the stability of the OER electrocatalysts, dopants have been introduced to RuO_x to modify the structure of the catalysts. Both DFT-based calculations and spectrometry analysis have proven that Ni doping is able to stabilise surface Ru and subsurface oxygen, preventing the deactivation of Ru-based materials [41].

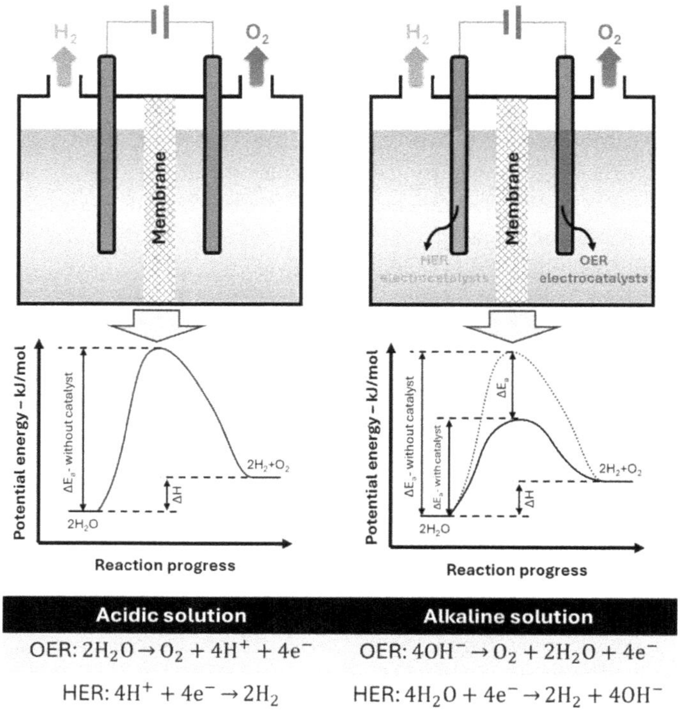

FIGURE 5.2 Promotion effect of electrocatalysts in water electrolysis [32,33].

5.2.1.2 Anion Exchange Membrane (AEM) Electrolysis

Anion exchange membrane (AEM) electrolysis is a promising alternative to PEM electrolysis. Unlike PEM, which operates in an acidic environment, AEM electrolysis operates under alkaline conditions. Similarly, electrocatalysts are required to facilitate the HER and OER and the transfer of the hydroxide ions between the cathode and the anode.

Platinum group metals (PGMs) are predominantly studied due to their excellent activity and good stability, with Pt and Ru being, once again, the most active material [42]. Extensive studies have also been carried out on Pt- and Ru-based alloys and phosphides, PtRu alloy has been shown to have higher power densities compared to Pt [43,44], and ruthenium phosphides have been shown to be active towards hydrogen evolution [45]. By alloying with transition metals, the *d*-band of Ru can be tuned to modify its electronic properties and enhance its catalytic performance. RuCr and RuCo were both found to be promising electrocatalysts due to their optimised binding energy towards hydroxide and hydrogen [46,47]. Pd-based materials, particularly when combined with CeO_2, exhibit high activity due to the oxophilic nature of ceria and the shift in the Pd-H desorption peak, leading to a lowered onsite potential [48,49]. Various PGM-free catalysts were developed by tuning the overall electronic structure and surface properties [50]. Nickel-based materials were alloyed with Mo [51] and Cu [52], or specially designed to form core–shell structures [53] and nitrides [54].

5.2.1.3 Alkaline Water Electrolysis (AWE)

Alkaline water electrolysis (AWE) is another well-established green hydrogen production method, in which a concentrated electrolyte solution of KOH or NaOH was used to split water electrochemically [55]. AWE stands out as a promising technology because of its high efficiency and relatively low cost due to the use of non-noble metal catalysts [56]. Although Pt- and Ir-/Ru-based materials and alloys were extensively studied as the HER and OER electrocatalysts [57], increasing attention has been paid to non-precious elements. Ni alloys, including NiAl, NiMo, and NiZn, were found to be suitable cathode catalysts, which was attributed to their corrosion resistance ability [58]. Other common anode materials include nickel oxides and ferrites (e.g., $NiFe_2O_4$ and $CoFe_2O_4$), thanks to their high oxidation's states and stability in the oxidising environment of the anode [59]. Apart from their composition, the morphology of the electrocatalysts also plays an important role in their overall performance. Hydride metallenes, which are atomically thin metal layers consisting of interstitial hydrogen atoms, show great promise in AEM and hydrogen-related catalysis processes [60]. A novel Ru-PdH_x metallene catalyst achieved exceptional HER activity in alkaline environment, with a low overpotential of 30 mV at 10 mA cm^{-2}, which surpassed commercial Pt/C and most Ru-based catalysts [61]. The performances of various electrocatalysts used in PEM, AEM, and AWE [62–65] were compared, and the result is shown in Figure 5.3. The high efficiency of an electrocatalyst used in electrolysis is reflected by a high mass density in the case of PEM, a high peak power density in AEM, and a low overpotential in AWE.

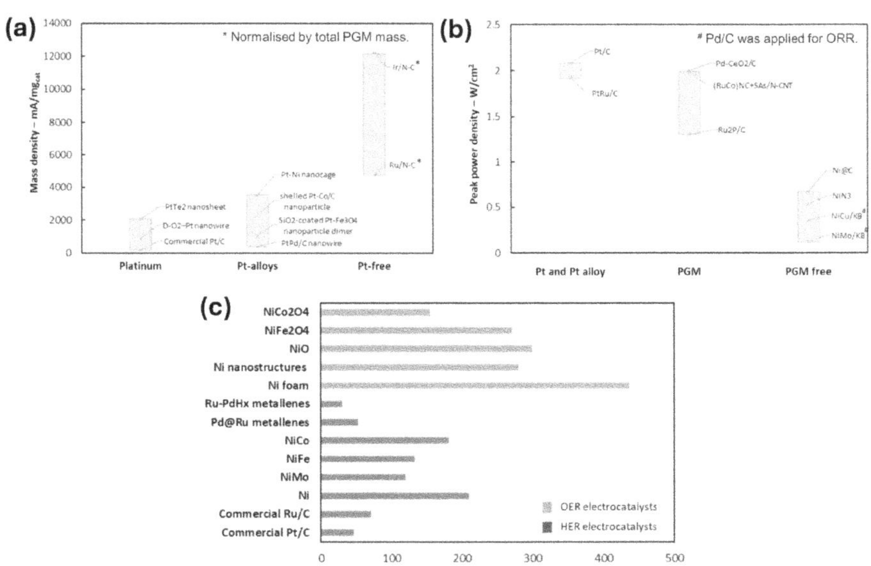

FIGURE 5.3 Comparison of the performance of the typical electrocatalysts in (a) PEM, (b) AEM, and (c) AWE.

5.2.2 Solid Looping Systems for 'Blue' Hydrogen Production

While PEM, AEM, and AWE provide carbon-free hydrogen through water-splitting powered by renewable energy, their large-scale deployment is often constrained. Blue hydrogen production technologies, such as sorption-enhanced steam methane reforming (SESMR) and chemical looping reforming (CLR), offer an alternative pathway by integrating carbon capture with conventional hydrogen production, effectively reducing CO_2 emissions.

5.2.2.1 Sorbent Enhanced Steam Methane Reforming

Sorption-enhanced steam methane reforming is an advanced hydrogen production process that integrates in situ CO_2 capture with traditional steam methane reforming. By using a solid sorbent, CO_2 is removed from the system as it forms, shifting the reaction equilibrium to the direction of higher hydrogen yield while reducing carbon emissions [66]. SESMR plays an important role in low-carbon hydrogen production, especially when integrated with renewable energy or biomass-derived methane, offering a scalable pathway to large-scale industrial use of blue hydrogen.

CaO-based sorbents are the most commonly studied sorbents for SESMR due to their high availability, good sorption capacity (78.6 wt.%), and fast sorption kinetics (0.1–0.4 g $CO_2 g^{-1} min^{-1}$) [67]. CaO-based sorbents can be derived from natural sources such as limestone or dolomite [68,69], or from industrial by-products such as steel slag [70–72]. One major challenge with CaO sorbents is their cyclic ability, as repeated calcination/carbonation cycles lead to CO_2 capture capacity loss and sintering [73]. Different synthesis techniques have been explored to enhance their performance, including organic acid treatment [74–76], thermal pretreatment [77], steam hydration [78], and the integration of thermally stable inert elements [79,80]. Additionally, maintaining a porous skeleton structure was also found to stabilise CaO sorbents [77]. Alkali metal-based sorbents, which typically contain alkali metal particles deposited on porous ceramic supports, capture CO_2 by forming stable carbonate compounds [81]. These materials exhibit superior mechanical stability, which makes them promising sorbents for long-term application. Binary-metal oxides, such as Li_2ZrO_3 and Na_2ZrO_3, were also able to capture CO_2 effectively at low concentrations and high temperatures [82]. Hydrotalcite-like compounds (or layered double hydroxides, LDHs) are also used as sorbents for SESMR due to their strong alkalinity and structural stability [83,84]. Composition modification by incorporating metal oxides was found to significantly improve the sorption capacity of hydrotalcites, achieving over 90% hydrogen purity [85–87]. Waste-derived sorbents, although studied less frequently, offer a cost-effective and sustainable alternative for CO_2 capture as these sorbents are usually produced from biomass, industrial by-products, and agricultural residues [88–90]. Repurposing such waste materials or by-products from other industries also aligns with the concept of circular economy, where waste is transformed into valuable inputs for low-carbon technologies. Bifunctional materials have also emerged as a promising approach for enhancing low-carbon hydrogen production. By combining highly active catalysts with CO_2 sorbents, such as $Ru/Ca_3Al_2O_6$–CaO, Co–CaO–CeO_2, and $Ca_9Al_6O_{18}$–CaO/NiO–CeO_2, high hydrogen purity and CO_2 capture capacity can be simultaneously achieved in multicycle processes [91–93].

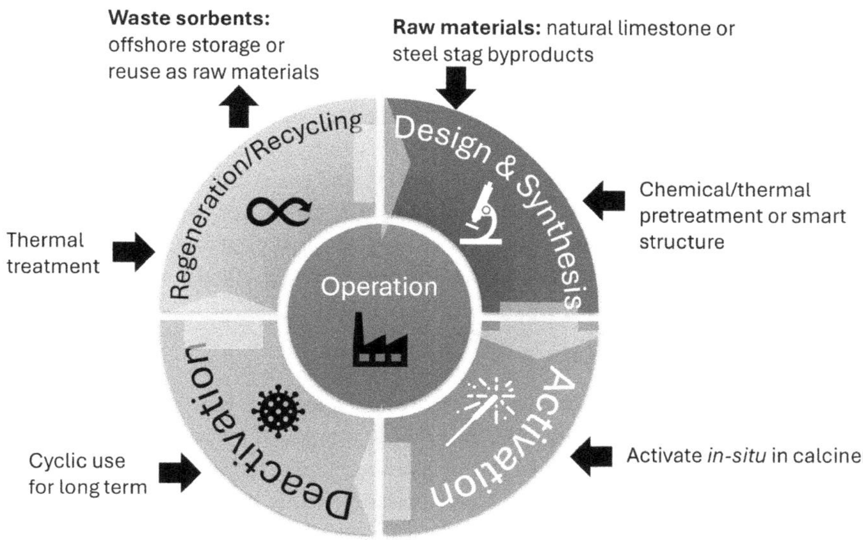

FIGURE 5.4 Life cycle of typical sorbents used in blue hydrogen production processes.

The life cycle of typical CaO-based sorbents used for SESMR and other low-carbon hydrogen production processes is summarised in Figure 5.4. The raw materials first undergo a chemical/thermal pretreatment process to remove any unwanted impurities and to enhance their absorptive properties. The sorbents are then activated and undergo long-term cyclic processes to capture CO_2. Finally, the saturated or deactivated sorbents are collected for offshore storage or regenerated and recycled for further use.

5.2.2.2 Chemical Looping (CL)

Chemical looping (CL) is a process in which a solid oxygen carrier (OC) is used to transfer oxygen between two interconnected reactors, enabling fuel combustion or conversion without direct contact with air. This technology significantly reduces carbon emissions by facilitating inherent CO_2 capture. CLR applies this principle to produce hydrogen or syngas by using an OC to partially oxidise hydrocarbons in a fuel reactor, followed by OC regeneration in an air reactor [94].

Research on suitable OC materials is the core of the CLR technology, with metal oxides being the most commonly used OCs. Iron oxide-based OC is the most studied material due to its high availability, low toxicity, and flexibility of operation due to its various chemical phases (Fe_2O_3, FeO, Fe_3O_4, etc.) [95]. However, one major challenge with iron oxide-based OCs is their low cyclability due to agglomeration and low reducibility [96]. Other metal oxide OCs include nickel oxide-based materials, which are known for their high catalytic activity for hydrocarbon reforming reactions and low attrition rate [97]. Copper and manganese oxides are also promising OC materials, due to their reactivity for complete oxidation of the fuel and oxygen uncoupling ability, as well as oxygen transport abilities [98]. Perovskite-based OCs have also been studied, as their unique crystalline structure allows for tailored redox properties by doping and partial cation substitution of the metal element sites [97]. For solar-driven CL processes, ceria was found to be a promising OC due to its high oxygen mobility and transport capacity at elevated temperatures [99]. Recent studies have also found waste or natural materials to be promising OC materials. Industrial wastes, such as red mud [100,101], slag [102,103], and electric arc furnace dust [104], as well as natural ores [105–107], were used for CL processes. Although their reactivity remains to be optimised, these materials provide a potential solution for controlling the cost of CL processes and the disposal pathway of wastes.

5.2.3 Hydrogen Storage

Effective hydrogen storage is essential for applications in fuel cells, grid energy storage, and transportation. Currently, the efficient storage of hydrogen remains a critical challenge. Hydrogen can be stored in gaseous, liquid, or solid-state forms, each with its advantages and limitations. Compressed gaseous hydrogen storage is commonly used but requires high pressures (several hundred bars), while liquid hydrogen requires cryogenic conditions (–253°C) [108]. Solid-state storage, using metal hydrides or porous materials, offers safer alternatives with higher volumetric density. In this section, recent advances in three pathways for hydrogen storage using advanced materials, including metal hydrides, porous materials, and ammonia, are summarised and critically reviewed. Overcoming current challenges in storage efficiency, cost, and safety will accelerate the transition towards a sustainable hydrogen economy.

5.2.3.1 Metal Hydrides

Metal hydrides store hydrogen through a reversible adsorption–desorption process, where hydrogen molecules are dissociated and incorporated into the metal lattice. The adsorption–desorption of hydrogen with transition metals can occur easily at elevated temperatures, and electrochemical approaches are usually used to evaluate the performance of the storage materials [109].

Pure magnesium or magnesium alloys are one of the most promising hydrogen storage materials. These materials are lightweight, abundant, and highly reversible [110]. Magnesium has a hydrogen storage capacity of approximately 7.7 wt.%, which is higher than those of compressed hydrogen (23 kgH m^{-3} at 350 bar) and liquid hydrogen (71 kgH m^{-3}) [111,112]. However, one of the major limitations related to the use of magnesium-based materials is the hydrogenation rate inhibition due to surface oxidation with air [113]. Methods to mitigate this issue include subjecting the material to heating/cooling cycles under hydrogen atmosphere and mechanical activation using ball mills [114]. Other commonly used metal hydride alloys include Mg_2Ni [115–117], TiFe [118], $TiMn_2$ [119,120], $LaNi_5$ [121,122], $NaAlH_4$ [123,124], and $LiBH_4$ [125]. Apart from the composition, other parameters also affect the storage efficiency of metal hydrides. Porous structures with high surface area and small particle size were found to be optimal for hydrogen uptake, which can be achieved through advanced synthesis methods, such as hydrolysis, sol–gel, combustion synthesis, high-energy ball milling, mechano-chemical activation, and melt spinning [110,126,127].

5.2.3.2 Porous Materials

Porous materials, with their unique characteristics, such as high tunability, large and accessible surface area, and high porosity, have stood out as promising hydrogen storage materials over the past decades [128]. Some of the most widely studied porous materials include MOFs, COFs, porous organic polymers (POPs), zeolites, and carbon-based materials.

The hydrogen storage capacity of MOFs depends on their pore size, volume, and surface area, with rational design and synthesis method crucial for optimisation. Highly porous MOFs, especially those with high-valent metal clusters (Zr_6, Al_3, Fe_3), exhibit good stability and hydrogen storage capacity due to their strong coordination bonds [129,130]. The hydrogen storage capacity of MOFs with optimised structures can reach as high as 50 g L^{-1} under cryogenic conditions [128], highlighting their potential as high-performance storage materials. Most MOFs interact relatively weakly with hydrogen under ambient temperature, requiring therefore cryogenic conditions for storage. Increasing charge density by increasing open metal sites and modifying pore structures were found to be effective methods to enhance adsorption strength [131–134]. MOFs with low-valent metal sites, which interact with hydrogen through π-back bonding, were also found to be promising candidates for hydrogen storage at ambient temperatures [135,136]. In addition to MOFs, COFs, with their crystalline structure and high surface area, are also able to achieve high hydrogen uptake, particularly when metal doping is introduced to enhance high-affinity sites [137–139]. POPs and zeolites, although offering good mechanical stability and well-defined microporous structures, generally exhibit moderate hydrogen uptake capacity and requires further optimisation in their physiochemical properties [128].

5.2.3.3 Ammonia

Ammonia is increasingly recognised as a promising hydrogen storage and transport medium due to its high hydrogen content, established infrastructure and global network, and mild storage conditions compared to gaseous/liquid hydrogen [140]. Once transported, ammonia can be used directly as a carbon-free fuel or cracked back into hydrogen. The Haber–Bosch process, which is the conventional pathway for ammonia synthesis, heavily relies on fossil fuels and is energy-intensive. However, recent developments in integrating renewable hydrogen with the Haber–Bosch process can significantly reduce its carbon footprints [141]. A more detailed description of the conversion between hydrogen and ammonia, as well as the advanced materials used for the decomposition and synthesis of ammonia, is covered in the 'Advanced Materials for Green Ammonia Production and Conversion' section.

5.2.4 Summary

Noble metals were widely applied in electrolysis process, while some transition metals, such as nickel, also played an important role in reducing the cost of catalysts for both 'green' and 'blue' hydrogen production. Novel sorbents and OCs were developed to accelerate the decarbonisation of 'blue' hydrogen production. However, it is still a challenge to store the hydrogen safely using hydrides or porous materials, and ammonia appears to be an alternative and effective option for hydrogen storage.

5.3 ADVANCED MATERIALS FOR GREEN AMMONIA PRODUCTION AND CONVERSION

Ammonia has emerged as an important component in the global effort to achieve the net-zero target. As a carbon-free species with high hydrogen content, ammonia is increasingly recognised as a promising hydrogen carrier and a sustainable energy vector. It can be synthesised using renewable sources, transported and stored efficiently, and decomposed to release hydrogen. Beyond energy, ammonia remains indispensable in industrial and agricultural applications, such as chemical synthesis and fertiliser production.

Catalysis plays an important role in the synthesis of ammonia and its conversion to hydrogen. In the Haber–Bosch process, which currently dominates global ammonia production, catalysts are essential for enabling the high-temperature and high-pressure reaction. Similarly, the cracking (or decomposition) of ammonia requires the use of catalysts to overcome kinetic barriers and achieve high conversion and product selectivity.

In this section, recent progress in the development of advanced materials for the main green ammonia synthesis techniques, including the Haber–Bosch process, electrochemical (EAS), photocatalytic (PAS), and biocatalytic ammonia synthesis (BAS); and for ammonia decomposition processes, including thermal cracking, electrochemical splitting, and plasma-assisted decomposition, is summarised and critically reviewed. The main focus is on the morphology and physicochemical properties of the supported catalysts, as well as the overall system design.

5.3.1 Advanced Catalysts for Green Ammonia Synthesis

Traditional ammonia synthesis relies on the Haber–Bosch process, which operates under high temperatures and high pressures to convert nitrogen and hydrogen into ammonia [142]. However, hydrogen used in the conventional Haber–Bosch process is generally derived from fossil fuel sources, predominantly hydrocarbon reforming, which leads to a significant amount of CO_2 emissions. Green ammonia synthesis focuses on alternative pathways, such as direct EAS, PAS, or integrating the conventional Haber–Bosch process with CCS technologies, using green hydrogen as the feedstock for the process. In this section, we focus on the application of advanced catalysts for the conventional Haber–Bosch process, as well as several novel green ammonia synthesis processes, namely electrochemical, photocatalytic, and biocatalytic ammonia synthesis.

5.3.1.1 Catalysts for the Haber–Bosch Process

In a typical ammonia synthesis reactor, magnetite iron catalysts are usually used under temperatures above 400°C and pressures around 150 bar [142]. However, the single-pass conversion under such conditions is quite low, usually around 20%. Wustite catalyst has been found to be a better option, lowering reaction pressure to 100 bar. The limitation, however, remains in its high recycling and feed compression costs [143]. Ru-based catalysts have also been studied as alternative ammonia synthesis catalysts. Although they have been proven to be active under atmospheric pressure and lower temperatures (below 400°C) [144], which are significantly milder conditions compared to their iron-based counterparts, Ru-based catalysts only account for less than 5% of global ammonia production due to their cost and scarcity [145]. One way to decarbonise the traditional Haber–Bosch process is by replacing hydrogen produced from methane reforming with hydrogen from electrolysis and powering equipment with electricity, rather than steam [142]. In this sense, advanced catalysts developed for hydrogen production from electrolysis will play an essential role in the novel, decarbonised electric Haber–Bosch process.

5.3.1.2 Catalysts for Electrochemical Ammonia Synthesis

The EAS process consists of the following reaction, which takes place at an electrode potential of 0.092 V at ambient temperature [146]:

$$N_2 + 6e^- + 6H^+ \rightarrow 2NH_3.$$

Compared to the conventional Haber–Bosch process, the advantages of EAS include mild reaction conditions, the possibility of integrating renewable electricity, fewer infrastructure requirements, and abundant carbon-free feedstock. Catalysts used for EAS can be divided into three categories: noble metals, non-noble metals, and non-metal elements. Ru, Rh, Au, and Pd are commonly used noble metal-based catalysts for EAS and have demonstrated good activity [147]. Efforts have also been made to minimise the cost associated with the use of noble metal catalysts. Single-atom catalysts, bimetallic catalysts, and catalysts with specially designed nanostructures, have been gaining increasing attention over the past decades. In particular, nanoporous palladium hydride [148], PdCu loaded on carbon powder [149], and noble metal nanocrystals [150] have shown great promise as advanced catalysts for the electrocatalytic synthesis of ammonia.

5.3.1.3 Catalysts for Photocatalytic and Biocatalytic Ammonia Synthesis

PAS is an emerging zero-carbon technology in which solar energy is harvested to produce ammonia. Photocatalysts are, in general, semiconductors with an electron-filled valence band and an electron-free conduction band, separated by a band gap. When exposed to sunlight, electrons in the valence band absorb photons and are excited to the conduction band. In PAS, electrons reduce N_2 to NH_3, while oxidising water to O_2 at the same time [146]. Metal oxides, TiO_2 in particular, are some of the most commonly used photocatalysts [151]. Internal modifications, such as increasing the concentration of the active Ti^{3+} active sites and constructing oxygen vacancies, have been made to conventional TiO_2 catalysts to boost the transfer of photogenerated electrons. On the other hand, external modification, such as doping additional elements (e.g., boron) into TiO_2 [152], was also found to be an efficient method to narrow the band gap and ease the excitation of the electrons. Bismuth oxyhalides ($Bi_xO_yM_z$, where M = Cl, Br, or I) are ternary semiconductors with layered structures, which are cost-effective, chemically stable, and highly selective to N_2 [153]. However, the performance of bismuth oxyhalides is hindered by their limited light absorption and charge separation capability. One solution to address these limitations is defect engineering, such as the preparation of single-cell-thick nanosheets [154] or nanotubes [155]. Other photocatalysts used for PAS include graphitic carbon nitride, which exhibits strong visible light absorption and is environmentally friendly as well as organic frameworks, which are attractive options due to their porous and adjustable structure [156,157].

In nature, approximately 48% of nitrogen is fixed by bacteria [146]. BAS involves creating artificial enzymes with the same nitrogenase activity as the naturally occurring nitrogen-fixing bacteria. The microbial nitrogen fixation reaction requires the transfer of 8 electrons and 16 magnesium adenosine triphosphate (ATP) to proceed, and forms NH_3, H_2, adenosine diphosphate (ADP), and phosphate from inorganic sources (Pi) [158]:

$$N_2 + 16Mg\,ATP + 8e^- + 8H^+ \rightarrow 2NH_3 + H_2 + 16Mg\,ADP + 16Pi.$$

It has been found that reduced ferredoxin and flavin redox proteins, which contain Mo and Fe, play an essential role in microbial nitrogen fixation [146]. Inspired by nature, Mo has been recognised as a promising active element for ammonia synthesis. Numerical studies based on first-principles calculations have demonstrated supported Mo-based catalysts to be active materials for nitrogen fixation [159], [160], exhibiting very low overpotential. A polymer network containing artificial groups of MoFe and Fe has also been tested [161], which identified Fe as an active site for N_2 reduction. While BAS is a promising technology, it is still a new field of study. A better understanding of the reaction mechanism and further research into more effective catalysts under mild conditions remain necessary.

5.3.2 Advanced Catalysts for Ammonia Decomposition/Conversion

Apart from its role as a fertiliser and a precursor for producing chemicals such as nitric acid and urea, ammonia is increasingly recognised as a promising hydrogen carrier. The advantages of using ammonia as a hydrogen carrier include the following:

- **High hydrogen content**: Ammonia has a hydrogen content of 17.6% by weight, which is higher than other potential hydrogen carriers, such as methane, methanol, and propane.
- **High volumetric density**: Ammonia offers a higher volumetric energy density compared to compressed hydrogen [162], making it an efficient medium for mid- to long-distance transport and storage.
- **Mild conditions for liquefaction**: Ammonia can be liquefied at –34°C at 1 bar or 25°C at 10 bar [163], making it suitable for applications in trains, airplanes, or ships.

Hydrogen can be produced from ammonia through several pathways, including thermal cracking [164], electrochemical splitting [165], and catalytic and plasma-assisted decomposition [166]. In this section, recent progress in the development of advanced catalysts is summarised and reviewed, with a focus on catalytic material, support design, and performance evaluation. A case study is also provided in this study to demonstrate an example of the industrial-scale production of green hydrogen through ammonia cracking using the advanced Metal Membrane Technology (MMT), which is currently under development by a consortium led by Siemens.

5.3.2.1 Catalysts for Ammonia Cracking

Ammonia cracking involves breaking down ammonia molecules into N_2 and H_2 at elevated temperatures, usually between 400°C and 700°C in the presence of a catalyst. Materials that can be used to activate this reaction include supported transition metals, metal alloys, and less commonly studied materials such as metal oxide hydrides, metal carbides, nitrides, imides, or amides [167,168]. Among the transition metals, Ru, Ni, Fe, and Co are some of the most studied catalytic materials. Despite the high cost, it is generally agreed that Ru-based materials are the most active catalysts for ammonia cracking, particularly at lower pressures [169]. Using the conventional and widely available γ-alumina as the support, Ru-Al_2O_3 was able to achieve a very high ammonia conversion (>90%) at 1 atm, with a Ru loading of less than 5% and at temperatures between 400°C and 500°C [170]. The excellent catalytic activity of Ru was attributed to the B5-type surface sites, which are active for the rate-determining step of N association and N-metal bond formation [171,172]. As B5-type sites are usually present in Ru nanoparticles with a diameter of approximately 2 nm, tailoring the particle size of the catalysts using advanced synthesis methods can greatly improve the overall performance of the Ru catalysts [173,174]. Apart from the particle size, the type of support also affects the overall performance of the Ru catalysts [175,176]. Metal oxides such as MgO [177,178], SiO_2 [179,180], ZrO_2 [181], TiO_2 [182], CeO_2 [183], and carbon nanotubes [184] were found to be optimal support materials for Ru-based catalysts. Nickel is also a promising candidate for ammonia cracking due to its abundance and good activity, comparable with Ru catalysts. In addition to conventional metal oxide supports, rare earth element-based supports have been proven to have particularly good synergy with Ni, with La_2O_3 [185], CeO_2 [186,187], Gd_2O_3 [188,189], and Sm_2O_3 [188,189] supported Ni catalysts achieving high ammonia conversion.

A recent collaboration led by Siemens Energy is constructing a prototype ammonia cracker for the production of green hydrogen at scale. The project aims to produce 200 kg hydrogen daily using the advanced MMT developed by Fortescue Future Industries [190]. The Metal Membrane, consisting of Pd-based catalysts for cracking ammonia and a vanadium or vanadium-alloy core, acts as a purification unit to simultaneously crack ammonia and separate hydrogen from unwanted gases, including nitrogen and the unreacted ammonia. The ammonia cracker, integrated with the MMT unit allows for the production of high-purity hydrogen at the point of consumption to power hydrogen buses and trucks, or to generate off-grid electricity to replace diesel generators. If proven successful, the next step will be to scale up the unit for grid-scale power production and heating.

5.3.2.2 Catalysts for Ammonia Electrochemical Splitting

Ammonia electrochemical splitting, or electrochemical ammonia oxidation reaction (eAOR), offers a more energy-efficient alternative to traditional thermal catalytic processes due to its mild operating conditions. Compared to water electrolysis for hydrogen production, the electrolysis of ammonia to produce nitrogen and hydrogen requires a much lower external voltage, which is only around 0.06 V [191]. In this process, NH_3 and OH^- at the anode electrode are converted to N_2 and H_2O, while the H_2O at the cathode electrode is converted to H_2 and OH^- [192]. The design of the electrocatalysts is essential to overcome challenges such as high overpotential, poisoning, and high cost [193]. Pt-based catalysts are predominantly used for eAOR due to their excellent performance [194]. Based on experimental and theoretical studies on the reaction mechanism of eAOR, $*NH_x$ are reactive intermediates, while *N is poisonous [195–199]. The high activity of Pt is attributed to its low adsorption strength to *N and the stable adsorption of *NH [200]. Pt catalysts were able to achieve pure hydrogen generation with a high Faraday efficiency of over 80% [201,202], and further improvement can be made by controlling the morphology of the Pt particles [203], choosing different support materials [204], engineering the facets present in the Pt particles [205,206], and using advanced synthesis methods [207]. Bimetallic Pt-based catalysts have also been extensively studied. Pt-Ir alloy was found to be highly active for eAOR due to the synergy between Pt, which generates high current density, and Ir, which reduces overpotential [208–210]. To address the challenges faced by Pt-based electrocatalysts, including high cost and quick deactivation, research has been carried out on non-noble metal electrocatalysts, namely nickel and copper. Ni has been shown to have a similar onset potential as Pt [211], while bimetallic Ni-Cu catalysts, especially in the form of LDHs, exhibit high catalytic activity and good stability due to their unique structure and surfaces rich in hydroxyl groups [212–214].

5.3.2.3 Catalysts for Plasma-Assisted Decomposition

Plasma-assisted ammonia decomposition involves the use of plasma technologies to decompose ammonia for green hydrogen production. Although ammonia decomposition can be achieved by using plasma activation only, with the addition of a catalyst, the yield and selectivity in hydrogen can be significantly improved [215]. Plasma-assisted catalytic ammonia decomposition can be divided into two categories – in-plasma catalysis (IPC) and post-plasma catalysis (PPC). In the IPC process, ammonia decomposition can be activated with only the support material, such as γ-alumina and $MgAl_2O_4$, which have high dielectric constants [216]. Similar to thermal catalytic ammonia decomposition, Ru/Al_2O_3 was found to be a highly active catalyst [217]. Mo_2N, which is inactive for thermal ammonia decomposition, exhibited over 60% efficiency when combined with plasma discharge [218]. On the other hand, PPC is less studied. Lin et al. [219] tested a two-stage process, where the initial plasma decomposition of ammonia was followed by thermal catalytic cracking with a NiO/Al_2O_3 catalyst placed downstream from the plasma reactor. This configuration was capable of enhancing hydrogen production with the highest hydrogen content of approximately 35% achieved.

5.3.3 Summary

Although various studies have demonstrated significant growth potential in the application of EAS, PAS, BAS, thermal catalytic cracking, eAOR, and plasma-assisted decomposition as green ammonia and hydrogen production pathways, key challenges remain in the commercialisation of these technologies. The reliance on noble metals, such as Ru and Pt, significantly increases cost, hindering large-scale application. Shifting away from rare earth elements and noble metals toward earth-abundant alternatives (such as Ni, Fe, or Co) also lowers the environmental impact of the technologies and improves long-term availability of the catalysts. This is particularly important for the EAS and catalytic ammonia cracking processes, where the use of noble metal-based catalysts is currently dominant. Additionally, further validation is required for the long-term stability and scalability of non-noble metal-based materials under industrial conditions. Developing cost-effective synthesis methods and strategies to enhance the durability of the materials is crucial for the future employment of these advanced technologies. Overcoming these barriers will be essential to unlock the full potential of ammonia as a carbon-free hydrogen carrier in the net-zero economy.

REFERENCES

1. A. E. Baumann, D. A. Burns, B. Liu, and V. S. Thoi, "Metal-organic framework functionalization and design strategies for advanced electrochemical energy storage devices," *Commun Chem*, vol. 2, no. 1, pp. 1–14, Jul. 2019. doi: 10.1038/s42004-019-0184-6
2. L. Jiao, J. Y. R. Seow, W. S. Skinner, Z. U. Wang, and H. L. Jiang, "Metal–organic frameworks: structures and functional applications," *Mater Today*, vol. 27, pp. 43–68, Jul. 2019. doi: 10.1016/J.MATTOD.2018.10.038
3. D. A. Yang, H. Y. Cho, J. Kim, S. T. Yang, and W. S. Ahn, "CO_2 capture and conversion using Mg-MOF-74 prepared by a sonochemical method," *Energy Environ Sci*, vol. 5, no. 4, pp. 6465–6473, Mar. 2012. doi: 10.1039/C1EE02234B
4. T. M. McDonald et al., "Cooperative insertion of CO_2 in diamine-appended metal-organic frameworks," *Nature*, vol. 519, no. 7543, pp. 303–308, Mar. 2015. doi: 10.1038/nature14327
5. A. Asghar, N. Iqbal, T. Noor, B. M. Kariuki, L. Kidwell, and T. L. Easun, "Efficient electrochemical synthesis of a manganese-based metal–organic framework for H_2 and CO_2 uptake," *Green Chem*, vol. 23, no. 3, pp. 1220–1227, Feb. 2021. doi: 10.1039/D0GC03292A
6. A. Kazemi, F. Moghadaskhou, M. A. Pordsari, F. Manteghi, A. Tadjarodi, and A. Ghaemi, "Enhanced CO_2 capture potential of UiO-66-NH_2 synthesized by sonochemical method: experimental findings and performance evaluation," *Sci Rep*, vol. 13, no. 1, pp. 1–15, Nov. 2023. doi: 10.1038/s41598-023-47221-6
7. S. Mahajan and M. Lahtinen, "Recent progress in metal-organic frameworks (MOFs) for CO_2 capture at different pressures," *J Environ Chem Eng*, vol. 10, no. 6, p. 108930, Dec. 2022. doi: 10.1016/J.JECE.2022.108930
8. J. Wang et al., "Hydration energetics of a diamine-appended metal–organic framework carbon capture sorbent," *J Phys Chem* C, vol. 124, no. 1, pp. 398–403, Jan. 2020. doi: 10.1021/ACS.JPCC.9B08008/ASSET/IMAGES/MEDIUM/JP9B08008_0002.GIF
9. C. Xiao, J. Tian, Q. Chen, and M. Hong, "Water-stable metal–organic frameworks (MOFs): rational construction and carbon dioxide capture," *Chem Sci*, vol. 15, no. 5, pp. 1570–1610, Jan. 2024. doi: 10.1039/D3SC06076D
10. R. P. Loughran et al., "CO_2 capture from wet flue gas using a water-stable and cost-effective metal-organic framework," *Cell Rep Phys Sci*, vol. 4, no. 7, p. 101470, Jul. 2023. doi: 10.1016/J.XCRP.2023.101470
11. A. M. Fracaroli et al., "Metal-organic frameworks with precisely designed interior for carbon dioxide capture in the presence of water," *J Am Chem Soc*, vol. 136, no. 25, pp. 8863–8866, Jun. 2014. doi: 10.1021/JA503296C/SUPPL_FILE/JA503296C_SI_003.CIF

12. H. C. Gulbalkan, G. O. Aksu, G. Ercakir, and S. Keskin, "accelerated discovery of metal–organic frameworks for CO_2 capture by artificial intelligence," *Ind Eng Chem Res*, vol. 63, no. 1, pp. 37–48, Jan. 2024. doi: 10.1021/ACS.IECR.3C03817/ASSET/IMAGES/LARGE/IE3C03817_0006.JPEG
13. Z. Yao et al., "Inverse design of nanoporous crystalline reticular materials with deep generative models," *Nat Mach Intell*, vol. 3, no. 1, pp. 76–86, Jan. 2021. doi: 10.1038/s42256-020-00271-1
14. H. Park et al., "A generative artificial intelligence framework based on a molecular diffusion model for the design of metal-organic frameworks for carbon capture," *Commun Chem*, vol. 7, no. 1, pp. 1–18, Feb. 2024. doi: 10.1038/s42004-023-01090-2
15. H. R. Abuzeid, A. F. M. EL- Mahdy, and S. W. Kuo, "Covalent organic frameworks: design principles, synthetic strategies, and diverse applications," *Giant*, vol. 6, p. 100054, Jun. 2021. doi: 10.1016/J.GIANT.2021.100054
16. H. Furukawa and O. M. Yaghi, "Storage of hydrogen, methane, and carbon dioxide in highly porous covalent organic frameworks for clean energy applications," *J Am Chem Soc*, vol. 131, no. 25, pp. 8875–8883, Jul. 2009. doi: 10.1021/JA9015765/SUPPL_FILE/JA9015765_SI_001.PDF
17. Y. Zhao, K. X. Yao, B. Teng, T. Zhang, and Y. Han, "A perfluorinated covalent triazine-based framework for highly selective and water–tolerant CO_2 capture," *Energy Environ Sci*, vol. 6, no. 12, pp. 3684–3692, Nov. 2013. doi: 10.1039/C3EE42548G
18. D. Kaleeswaran, P. Vishnoi, and R. Murugavel, "[3+3] Imine and β-ketoenamine tethered fluorescent covalent-organic frameworks for CO_2 uptake and nitroaromatic sensing," *J Mater Chem C Mater*, vol. 3, no. 27, pp. 7159–7171, Jul. 2015. doi: 10.1039/C5TC00670H
19. Z. Zhou et al., "Carbon dioxide capture from open air using covalent organic frameworks," *Nature*, vol. 635, no. 8037, pp. 96–101, Oct. 2024. doi: 10.1038/s41586-024-08080-x
20. Y. Ge et al., "Understanding water adsorption and the impact on CO_2 capture in chemically stable covalent organic frameworks," *J Phys Chem* C, vol. 122, no. 48, pp. 27495–27506, Dec. 2018. doi: 10.1021/ACS.JPCC.8B09033/SUPPL_FILE/JP8B09033_SI_001.PDF
21. Q. Qu et al., "Screening and preparation of functionalized TpBD-COFs for CO_2 capture," *Chem Eng Sci*, vol. 301, p. 120702, Jan. 2025. doi: 10.1016/J.CES.2024.120702
22. J. S. De Vos et al., "High-throughput screening of covalent organic frameworks for carbon capture using machine learning," *Chem Mater*, vol. 36, no. 9, pp. 4315–4330, May 2024. doi: 10.1021/ACS.CHEMMATER.3C03230/ASSET/IMAGES/LARGE/CM3C03230_0007.JPEG
23. K. S. Deeg et al., "In silico discovery of covalent organic frameworks for carbon capture," *ACS Appl Mater Interfaces*, vol. 12, no. 19, pp. 21559–21568, May 2020. doi: 10.1021/ACSAMI.0C01659/SUPPL_FILE/AM0C01659_SI_002.PDF
24. H. Li, A. Dilipkumar, S. Abubakar, and D. Zhao, "Covalent organic frameworks for CO_2 capture: from laboratory curiosity to industry implementation," *Chem Soc Rev*, vol. 52, no. 18, pp. 6294–6329, Sep. 2023. doi: 10.1039/D2CS00465H
25. S. K. Gebremariam, L. F. Dumée, P. L. Llewellyn, Y. F. Alwahedi, and G. N. Karanikolos, "Metal-organic framework hybrid adsorbents for carbon capture – a review," *J Environ Chem Eng*, vol. 11, no. 2, p. 109291, Apr. 2023. doi: 10.1016/J.JECE.2023.109291
26. T. Rodenas et al., "Metal–organic framework nanosheets in polymer composite materials for gas separation," *Nat Mater 2014* 14:1, vol. 14, no. 1, pp. 48–55, Nov. 2014. doi: 10.1038/nmat4113
27. F. Yang, J. Wu, X. Zhu, T. Ge, and R. Wang, "Enhanced stability and hydrophobicity of LiX@ZIF-8 composite synthesized environmental friendly for CO_2 capture in highly humid flue gas," *Chem Eng J*, vol. 410, p. 128322, Apr. 2021. doi: 10.1016/J.CEJ.2020.128322

28. B. Seoane et al., "Metal–organic framework based mixed matrix membranes: a solution for highly efficient CO_2 capture?," *Chem Soc Rev*, vol. 44, no. 8, pp. 2421–2454, Apr. 2015. doi: 10.1039/C4CS00437J
29. Y. Cheng et al., "Mixed matrix membranes containing MOF@COF hybrid fillers for efficient CO_2/CH_4 separation," *J Memb Sci*, vol. 573, pp. 97–106, Mar. 2019. doi: 10.1016/J.MEMSCI.2018.11.060
30. L. F. Feitosa et al., "Surface molecular design of organic–inorganic mesoporous hybrid materials for CO_2 capture," *J Environ Chem Eng*, vol. 9, no. 1, p. 104951, Feb. 2021. doi: 10.1016/J.JECE.2020.104951
31. N. Sezer, S. Bayhan, U. Fesli, and A. Sanfilippo, "A comprehensive review of the state-of-the-art of proton exchange membrane water electrolysis," *Mater Sci Energy Technol*, vol. 8, pp. 44–65, 2025. doi: 10.1016/j.mset.2024.07.006
32. A. Kazemi, F. Manteghi, and Z. Tehrani, "Metal electrocatalysts for hydrogen production in water splitting," ACS Omega, Jan. 2024. doi: 10.1021/acsomega.3c07911
33. U. Abdullah, M. Ali, and E. Pervaiz, "An inclusive review on recent advancements of cadmium sulfide nanostructures and its hybrids for photocatalytic and electrocatalytic applications," *Mol Catal*, vol. 508, p. 111575, May 2021. doi: 10.1016/j.mcat.2021.111575
34. X. Tian et al., "Engineering bunched Pt-Ni alloy nanocages for efficient oxygen reduction in practical fuel cells," *Science*, vol. 366, no. 6467, pp. 850–856, Nov. 2019. doi: 10.1126/science.aaw7493
35. T. Duc Le et al., "Synergistic effect of Pt-Ni dual single-atoms and alloy nanoparticles as a high-efficiency electrocatalyst to minimize Pt utilization at cathode in polymer electrolyte membrane fuel cells," *J Colloid Interface Sci*, vol. 634, pp. 930–939, Mar. 2023. doi: 10.1016/j.jcis.2022.12.061
36. L. Chong et al., "Highly durable fuel cell electrocatalyst with low-loading Pt-Co nanoparticles dispersed over single-atom Pt-Co-N-graphene nanofiber," *Chem Catalysis*, vol. 3, no. 3, p. 100541, Mar. 2023. doi: 10.1016/j.checat.2023.100541
37. B. Yang, X. Yu, J. Hou, and Z. Xiang, "Secondary reduction strategy synthesis of Pt–Co nanoparticle catalysts towards boosting the activity of proton exchange membrane fuel cells," *Particuology*, vol. 79, pp. 18–26, Aug. 2023. doi: 10.1016/j.partic.2022.11.010
38. X. Kong et al., "Advanced materials and fabricating approaches for proton exchange membrane fuel cells," *Responsive Mater*, vol. 2, no. 4, Nov. 2024. doi: 10.1002/rpm.20240028
39. X. Deng et al., "Recent progress in materials design and fabrication techniques for membrane electrode assembly in proton exchange membrane fuel cells," *Catalysts*, vol. 15, no. 1, p. 74, Jan. 2025. doi: 10.3390/catal15010074
40. S. Siracusano, N. Van Dijk, E. Payne-Johnson, V. Baglio, and A. S. Aricò, "Nanosized IrO_x and $IrRuO_x$ electrocatalysts for the O_2 evolution reaction in PEM water electrolysers," *Appl Catal B*, vol. 164, pp. 488–495, Mar. 2015. doi: 10.1016/j.apcatb.2014.09.005
41. Z.-Y. Wu et al., "Non-iridium-based electrocatalyst for durable acidic oxygen evolution reaction in proton exchange membrane water electrolysis," *Nat Mater*, vol. 22, no. 1, pp. 100–108, Jan. 2023. doi: 10.1038/s41563-022-01380-5
42. H. Lei et al., "Multiscale understanding of anion exchange membrane fuel cells: mechanisms, electrocatalysts, polymers, and cell management," *Adv Mater*, Jan. 2025. doi: 10.1002/adma.202410106
43. W. Ni et al., "Synergistic interactions between PtRu catalyst and nitrogen-doped carbon support boost hydrogen oxidation," *Nat Catal*, vol. 6, no. 9, pp. 773–783, Aug. 2023. doi: 10.1038/s41929-023-01007-1
44. Y. Wang et al., "Pt–Ru catalyzed hydrogen oxidation in alkaline media: oxophilic effect or electronic effect?," *Energy Environ Sci*, vol. 8, no. 1, pp. 177–181, 2015. doi: 10.1039/C4EE02564D

45. Y. Zhao et al., "High-performance Ru_2 P anodic catalyst for alkaline polymer electrolyte fuel cells," CCS Chem, vol. 4, no. 5, pp. 1732–1744, May 2022. doi: 10.31635/ccschem.021.202100810
46. Z. Cui et al., "Dilute RuCo alloy synergizing single Ru and Co atoms as efficient and CO-resistant anode catalyst for anion exchange membrane fuel cells," *Angew Chem Int Ed*, vol. 63, no. 28, Jul. 2024. doi: 10.1002/anie.202404761
47. C. Yang et al., "The role of hydroxide binding energy in alkaline hydrogen oxidation reaction kinetics on RuCr NANOSHEET†," *Chin J Chem*, vol. 40, no. 21, pp. 2495–2501, Nov. 2022. doi: 10.1002/cjoc.202200385
48. Z. A. Feng, F. El Gabaly, X. Ye, Z.-X. Shen, and W. C. Chueh, "Fast vacancy-mediated oxygen ion incorporation across the ceria–gas electrochemical interface," *Nat Commun*, vol. 5, no. 1, p. 4374, Jul. 2014. doi: 10.1038/ncomms5374
49. H. A. Miller et al., "A Pd/C-CeO_2 anode catalyst for high-performance platinum-free anion exchange membrane fuel cells," *Angew Chem Int Ed*, vol. 55, no. 20, pp. 6004–6007, May 2016. doi: 10.1002/anie.201600647
50. S. Lu, J. Pan, A. Huang, L. Zhuang, and J. Lu, "Alkaline polymer electrolyte fuel cells completely free from noble metal catalysts," *Proc Natl Acad Sci USA*, vol. 105, no. 52, pp. 20611–20614, Dec. 2008. doi: 10.1073/pnas.0810041106
51. S. Kabir et al., "Platinum group metal-free NiMo hydrogen oxidation catalysts: high performance and durability in alkaline exchange membrane fuel cells," *J Mater Chem A Mater*, vol. 5, no. 46, pp. 24433–24443, 2017. doi: 10.1039/C7TA08718G
52. G. Wang, W. Li, B. Huang, L. Xiao, J. Lu, and L. Zhuang, "Exploring the composition–activity relation of Ni–Cu binary alloy electrocatalysts for hydrogen oxidation reaction in alkaline media," *ACS Appl Energy Mater*, vol. 2, no. 5, pp. 3160–3165, May 2019. doi: 10.1021/acsaem.8b02206
53. L. Gao et al., "A nickel nanocatalyst within a h-BN shell for enhanced hydrogen oxidation reactions," *Chem Sci*, vol. 8, no. 8, pp. 5728–5734, 2017. doi: 10.1039/C7SC01615H
54. Z. Cui, R. G. Burns, and F. J. DiSalvo, "Mesoporous $Ti_{0.5}$ $Nb_{0.5}N$ ternary nitride as a novel noncarbon support for oxygen reduction reaction in acid and alkaline electrolytes," *Chem Mater*, vol. 25, no. 19, pp. 3782–3784, Oct. 2013. doi: 10.1021/cm4027545
55. Y. Wu, G. Xu, J. Zhou, and D. Cao, "Research progress of the porous membranes in alkaline water electrolysis for green hydrogen production," *Chem Eng J*, vol. 505, p. 159291, Feb. 2025. doi: 10.1016/j.cej.2025.159291
56. S. Dash et al., "Advances in green hydrogen production through alkaline water electrolysis: a comprehensive review," *Int J Hydrogen Energy*, vol. 83, pp. 614–629, Sep. 2024. doi: 10.1016/j.ijhydene.2024.08.157
57. G. Gao et al., "Recent advancements in noble-metal electrocatalysts for alkaline hydrogen evolution reaction," *Chin Chem Lett*, vol. 36, no. 1, p. 109557, Jan. 2025. doi: 10.1016/j.cclet.2024.109557
58. A. L. Santos, M.-J. Cebola, and D. M. F. Santos, "Towards the hydrogen economy – a review of the parameters that influence the efficiency of alkaline water electrolyzers," *Energies*, vol. 14, no. 11, p. 3193, May 2021. doi: 10.3390/en14113193
59. D. Chanda, J. Hnát, T. Bystron, M. Paidar, and K. Bouzek, "Optimization of synthesis of the nickel-cobalt oxide based anode electrocatalyst and of the related membrane-electrode assembly for alkaline water electrolysis," *J Power Sources*, vol. 347, pp. 247–258, Apr. 2017. doi: 10.1016/j.jpowsour.2017.02.057
60. J. Fan et al., "Hydrogen stabilized RhPdH 2D bimetallene nanosheets for efficient alkaline hydrogen evolution," *J Am Chem Soc*, vol. 142, no. 7, pp. 3645–3651, Feb. 2020. doi: 10.1021/jacs.0c00218
61. J. Fan et al., "Spatially confined PdH_x metallenes by tensile strained atomic Ru layers for efficient hydrogen evolution," *J Am Chem Soc*, vol. 145, no. 10, pp. 5710–5717, Mar. 2023. doi: 10.1021/jacs.2c11692

62. J. Fan et al., "Spatially CONFINED PdHx metallenes by tensile strained atomic Ru layers for efficient hydrogen evolution," *J Am Chem Soc*, vol. 145, no. 10, pp. 5710–5717, Mar. 2023. doi: 10.1021/JACS.2C11692/ASSET/IMAGES/LARGE/JA2C11692_0006.JPEG
63. L. Mølmen, K. Eiler, L. Fast, P. Leisner, and E. Pellicer, "Recent advances in catalyst materials for proton exchange membrane fuel cells," *APL Mater*, vol. 9, no. 4, p. 40702, Apr. 2021. doi: 10.1063/5.0045801/123030
64. H. Lei et al., "Multiscale understanding of anion exchange membrane fuel cells: mechanisms, electrocatalysts, polymers, and cell management," *Adv Mater*, vol. 37, no. 8, p. 2410106, Feb. 2025. doi: 10.1002/ADMA.202410106
65. I. M. Perović, S. D. Mitrović, S. M. Brković, and I. A. Pašti, "Advances in nickel-based catalysts for alkaline water electrolysis: comprehensive review of current research direction for HER and OER Applications," Chemical Record, vol. 2025, p. e202500049, 2025. doi: 10.1002/TCR.202500049
66. S. Wang, Z. Shen, A. Osatiashtiani, A. Nabavi, and P. Clough, "Ni-based bimetallic catalysts for hydrogen production via (sorption-enhanced) steam methane reforming," https://ssrn.com/abstract=4635693 or doi.: 10.2139/ssrn.4635693
67. G. Ji et al., "Recent advances on kinetics of carbon dioxide capture using solid sorbents at elevated temperatures," *Appl Energy*, vol. 267, p. 114874, Jun. 2020. doi: 10.1016/j.apenergy.2020.114874
68. K. Wang, X. Hu, P. Zhao, and Z. Yin, "Natural dolomite modified with carbon coating for cyclic high-temperature CO_2 capture," *Appl Energy*, vol. 165, pp. 14–21, Mar. 2016. doi: 10.1016/j.apenergy.2015.12.071
69. H. Gupta and L.-S. Fan, "Carbonation–calcination cycle using high reactivity calcium oxide for carbon dioxide separation from flue gas," *Ind Eng Chem Res*, vol. 41, no. 16, pp. 4035–4042, Aug. 2002. doi: 10.1021/ie0108671
70. Y. Li, R. Sun, C. Liu, H. Liu, and C. Lu, "CO_2 capture by carbide slag from chlor-alkali plant in calcination/carbonation cycles," *Int J Greenh Gas Control*, vol. 9, pp. 117–123, Jul. 2012. doi: 10.1016/j.ijggc.2012.03.012
71. Y. Li, C. Liu, R. Sun, H. Liu, S. Wu, and C. Lu, "Sequential SO_2 / CO_2 capture of calcium-based solid waste from the paper industry in the calcium looping process," *Ind Eng Chem Res*, vol. 51, no. 49, pp. 16042–16048, Dec. 2012. doi: 10.1021/ie301375g
72. B. Guo et al., "Synthesis of vaterite $CaCO_3$ micro-spheres by carbide slag and a novel CO_2-storage material," *J* $CO_{2\ UtilCO2}$, vol. 18, pp. 23–29, Mar. 2017. doi: 10.1016/j.jcou.2017.01.004
73. A. N. Antzaras and A. A. Lemonidou, "Recent advances on materials and processes for intensified production of blue hydrogen," *Renew Sustain Energy Rev*, vol. 155, p. 111917, Mar. 2022. doi: 10.1016/j.rser.2021.111917
74. Y. Li et al., "Cyclic calcination/carbonation looping of dolomite modified with acetic acid for $CO_{2\ capture,}$" Fuel Process Technol, vol. 89, no. 12, pp. 1461–1469, Dec. 2008. doi: 10.1016/j.fuproc.2008.07.008
75. F. N. Ridha, V. Manovic, A. Macchi, M. A. Anthony, and E. J. Anthony, "Assessment of limestone treatment with organic acids for CO_2 capture in Ca-looping cycles," *Fuel Process Technol*, vol. 116, pp. 284–291, Dec. 2013. doi: 10.1016/j.fuproc.2013.07.007
76. F. N. Ridha, V. Manovic, Y. Wu, A. Macchi, and E. J. Anthony, "Post-combustion CO_2 capture by formic acid-modified CaO-based sorbents," *Int J Greenh Gas Control*, vol. 16, pp. 21–28, Aug. 2013. doi: 10.1016/j.ijggc.2013.02.026
77. V. Manovic and E. J. Anthony, "Thermal activation of CaO-based sorbent and self-reactivation during CO_2 capture looping cycles," *Environ Sci Technol*, vol. 42, no. 11, pp. 4170–4174, Jun. 2008. doi: 10.1021/es800152s
78. M. Shokrollahi Yancheshmeh, H. R. Radfarnia, and M. C. Iliuta, "Influence of steam addition during carbonation or calcination on the CO_2 capture performance of $Ca_9Al_6O_{18}CaO$ sorbent," *J Nat Gas Sci Eng*, vol. 36, pp. 1062–1069, Nov. 2016. doi: 10.1016/j.jngse.2016.06.009

79. S. D. Angeli, C. S. Martavaltzi, and A. A. Lemonidou, "Development of a novel-synthesized Ca-based CO_2 sorbent for multicycle operation: parametric study of sorption," *Fuel*, vol. 127, pp. 62–69, Jul. 2014. doi: 10.1016/j.fuel.2013.10.046
80. C. S. Martavaltzi and A. A. Lemonidou, "Development of new CaO based sorbent materials for CO_2 removal at high temperature," *Microporous Mesoporous Mater*, vol. 110, no. 1, pp. 119–127, Apr. 2008. doi: 10.1016/j.micromeso.2007.10.006
81. A. S. Farooqi et al., "Advancements in sorption-enhanced steam reforming for clean hydrogen production: a comprehensive review," *Carbon Capture Sci Technol*, vol. 14, p. 100336, Mar. 2025. doi: 10.1016/j.ccst.2024.100336
82. B. Alcántar-Vázquez, E. Vera, F. Buitron-Cabrera, H. A. Lara-García, and H. Pfeiffer, "Evidence of CO oxidation–chemisorption process on sodium zirconate (Na_2ZrO_3)," *Chem Lett*, vol. 44, no. 4, pp. 480–482, Apr. 2015. doi: 10.1246/cl.141134
83. K. D. Dewoolkar and P. D. Vaidya, "Improved hydrogen production by sorption-enhanced steam methane reforming over hydrotalcite- and calcium-based hybrid materials," *Energy Fuels*, vol. 29, no. 6, pp. 3870–3878, Jun. 2015. doi: 10.1021/acs.energyfuels.5b00584
84. Z. Yong and A. E. Rodrigues, "Hydrotalcite-like compounds as adsorbents for carbon dioxide," *Energy Convers Manag*, vol. 43, no. 14, pp. 1865–1876, Sep. 2002. doi: 10.1016/S0196–8904(01)00125-X
85. K. D. Dewoolkar and P. D. Vaidya, "Sorption-enhanced steam reforming of ethylene glycol over dual functional hydrotalcite materials promoted with Pt and Ru," *ChemistrySelect*, vol. 2, no. 27, pp. 8326–8336, Sep. 2017. doi: 10.1002/slct.201701007
86. S. A. Ghungrud and P. D. Vaidya, "Sorption-enhanced reaction process for glycerol-to-hydrogen conversion over cobalt catalyst supported on promoted hydrotalcites," *Int J Hydrogen Energy*, vol. 45, no. 16, pp. 9440–9450, Mar. 2020. doi: 10.1016/j.ijhydene.2020.01.206
87. R. Dębek, M. Motak, M. E. Galvez, P. Da Costa, and T. Grzybek, "Catalytic activity of hydrotalcite-derived catalysts in the dry reforming of methane: on the effect of Ce promotion and feed gas composition," *React Kinet Mech Catal*, vol. 121, no. 1, pp. 185–208, Jun. 2017. doi: 10.1007/s11144-017-1167-1
88. N. Wang, Y. Feng, Y. Chen, and X. Guo, "Lithium-based sorbent from rice husk materials for hydrogen production via sorption-enhanced steam reforming of ethanol," *Fuel*, vol. 245, pp. 263–273, Jun. 2019. doi: 10.1016/j.fuel.2019.02.048
89. K. Gao, M. Shokrollahi Yancheshmeh, J. Duchesne, and M. C. Iliuta, "Valorization of coal fly ash as a stabilizer for the development of Ni/CaO-based bifunctional material," *ACS Sustain Chem Eng*, vol. 8, no. 9, pp. 3885–3895, Mar. 2020. doi: 10.1021/acssuschemeng.9b07360
90. F. Wang et al., "Enhancing hydrogen production from biomass pyrolysis by dental-wastes-derived sodium zirconate," *Int J Hydrogen Energy*, vol. 44, no. 43, pp. 23846–23855, Sep. 2019. doi: 10.1016/j.ijhydene.2019.07.095
91. M. Shokrollahi Yancheshmeh, H. R. Radfarnia, and M. C. Iliuta, "Sustainable production of high-purity hydrogen by sorption enhanced steam reforming of glycerol over CeO_2-promoted $Ca_9Al_6O_{18}$ –CaO/NiO bifunctional material," *ACS Sustain Chem Eng*, vol. 5, no. 11, pp. 9774–9786, Nov. 2017. doi: 10.1021/acssuschemeng.7b01627
92. S. A. Ghungrud and P. D. Vaidya, "Improved hydrogen production from Sorption-Enhanced Steam Reforming of Ethanol (SESRE) using multifunctional materials of cobalt catalyst and Mg-, Ce-, and Zr-modified CaO sorbents," *Ind Eng Chem Res*, vol. 59, no. 2, pp. 693–703, Jan. 2020. doi: 10.1021/acs.iecr.9b05472
93. S. M. Kim et al., "Bi-functional Ru/$Ca_3Al_2O_6$ –CaO catalyst- CO_2 sorbent for the production of high purity hydrogen *via* sorption-enhanced steam methane reforming," *Catal Sci Technol*, vol. 9, no. 20, pp. 5745–5756, 2019. doi: 10.1039/C9CY01095E

94. R. Ramezani, L. Di Felice, and F. Gallucci, "A review of chemical looping reforming technologies for hydrogen production: recent advances and future challenges," *J Phys*: Energy, vol. 5, no. 2, p. 024010, Apr. 2023. doi: 10.1088/2515–7655/acc4e8
95. Z. Yu et al., "Iron-based oxygen carriers in chemical looping conversions: a review," Carbon Res Conv, vol. 2, no. 1, pp. 23–34, Apr. 2019. doi: 10.1016/j.crcon.2018.11.004
96. J. Hu, S. Chen, and W. Xiang, "Sintering and agglomeration of Fe_2O_3-$MgAl_2O_4$ oxygen carriers with different Fe_2O_3 loadings in chemical looping processes," *Fuel*, vol. 265, p. 116983, Apr. 2020. doi: 10.1016/j.fuel.2019.116983
97. O. Palone, L. Cedola, D. Borello, and C. N. Markides, "Decarbonizing power and fuels production by chemical looping processes: systematic review and future perspectives," *Appl Therm Eng*, vol. 254, p. 123844, Oct. 2024. doi: 10.1016/j.applthermaleng.2024.123844
98. G. Liu and G. Lisak, "Cu-based oxygen carriers for chemical looping processes: opportunities and challenges," *Fuel*, vol. 342, p. 127828, Jun. 2023. doi: 10.1016/j.fuel.2023.127828
99. J. R. Fosheim, B. J. Hathaway, and J. H. Davidson, "High efficiency solar chemical-looping methane reforming with ceria in a fixed-bed reactor," *Energy*, vol. 169, pp. 597–612, Feb. 2019. doi: 10.1016/j.energy.2018.12.037
100. Z. Gu, L. Zhang, C. Lu, S. Qing, and K. Li, "Enhanced performance of copper ore oxygen carrier by red mud modification for chemical looping combustion," *Appl Energy*, vol. 277, p. 115590, Nov. 2020. doi: 10.1016/j.apenergy.2020.115590
101. Y. Wang, X. Tian, H. Zhao, and K. Liu, "The use of a low-cost oxygen carrier prepared from red mud and copper ore for in situ gasification chemical looping combustion of coal," *Fuel Process Technol*, vol. 205, p. 106460, Aug. 2020. doi: 10.1016/j.fuproc.2020.106460
102. M. Durmaz, N. Dilmaç, and Ö. F. Dilmaç, "Evaluation of performance of copper converter slag as oxygen carrier in chemical-looping combustion (CLC)," *Energy*, vol. 196, p. 117055, Apr. 2020. doi: 10.1016/j.energy.2020.117055
103. A. Hedayati, A. H. Soleimanisalim, T. Mattisson, and A. Lyngfelt, "Thermochemical conversion of biomass volatiles via chemical looping: comparison of ilmenite and steel converter waste materials as oxygen carriers," *Fuel*, vol. 313, p. 122638, Apr. 2022. doi: 10.1016/j.fuel.2021.122638
104. I. Wang et al., "A short-cut chemical looping hydrogen generation system by using iron-based material from steel industry," *Chem Eng J*, vol. 394, p. 124882, Aug. 2020. doi: 10.1016/j.cej.2020.124882
105. Z. Hu, Z. Miao, J. Wu, and E. Jiang, "Nickel-iron modified natural ore oxygen carriers for chemical looping steam methane reforming to produce hydrogen," *Int J Hydrogen Energy*, vol. 46, no. 80, pp. 39700–39718, Nov. 2021. doi: 10.1016/j.ijhydene.2021.09.242
106. Y. Tan, F. N. Ridha, M. A. Duchesne, D. Y. Lu, and R. W. Hughes, "Reduction kinetics of ilmenite ore as an oxygen carrier for pressurized chemical looping combustion of methane," *Energy Fuels*, vol. 31, no. 7, pp. 7598–7605, Jul. 2017. doi: 10.1021/acs.energyfuels.7b01038
107. C. Linderholm, M. Schmitz, P. Knutsson, and A. Lyngfelt, "Chemical-looping combustion in a 100-kW unit using a mixture of ilmenite and manganese ore as oxygen carrier," *Fuel*, vol. 166, pp. 533–542, Feb. 2016. doi: 10.1016/j.fuel.2015.11.015
108. N. Kumar, S.-Y. Lee, and S.-J. Park, "Advancements in hydrogen storage technologies: a comprehensive review of materials, methods, and economic policy," *Nano Today*, vol. 56, p. 102302, Jun. 2024. doi: 10.1016/j.nantod.2024.102302
109. A. Züttel, "Materials for hydrogen storage," *Mater Today*, vol. 6, no. 9, pp. 24–33, Sep. 2003. doi: 10.1016/S1369–7021(03)00922-2
110. K. S. Nivedhitha et al., "Advances in hydrogen storage with metal hydrides: mechanisms, materials, and challenges," *Int J Hydrogen Energy*, vol. 61, pp. 1259–1273, Apr. 2024. doi: 10.1016/j.ijhydene.2024.02.335

111. S. Gao et al., "Effects of nano-composites (FeB, FeB/CNTs) on hydrogen storage properties of MgH2," *J Power Sources*, vol. 438, p. 227006, Oct. 2019. doi: 10.1016/j.jpowsour.2019.227006
112. A. Zaluska, L. Zaluski, and J. O. Ström–Olsen, "Nanocrystalline magnesium for hydrogen storage," *J Alloys Compd*, vol. 288, no. 1–2, pp. 217–225, Jun. 1999. doi: 10.1016/S0925–8388(99)00073-0
113. N. Klopčič, I. Grimmer, F. Winkler, M. Sartory, and A. Trattner, "A review on metal hydride materials for hydrogen storage," *J Energy Storage*, vol. 72, p. 108456, Nov. 2023. doi: 10.1016/j.est.2023.108456
114. V. A. Dekhtyarenko, D. G. Savvakin, V. I. Bondarchuk, V. M. Shyvaniuk, T. V. Pryadko, and O. O. Stasiuk, "$TiMn_2$-based intermetallic alloys for hydrogen accumulation: problems and prospects," *Uspehi Fiziki Metallov*, vol. 22, no. 3, pp. 307–351, Aug. 2021. doi: 10.15407/ufm.22.03.307
115. D. Khan, J. Zou, X. Zeng, and W. Ding, "Hydrogen storage properties of nanocrystalline Mg_2Ni prepared from compressed $2MgH_2Ni$ powder," *Int J Hydrogen Energy*, vol. 43, no. 49, pp. 22391–22400, Dec. 2018. doi: 10.1016/j.ijhydene.2018.10.055
116. A. K. Singh, A. K. Singh, and O. N. Srivastava, "On the synthesis of the Mg_2Ni alloy by mechanical alloying," *J Alloys Compd*, vol. 227, no. 1, pp. 63–68, Aug. 1995. doi: 10.1016/0925–8388(95)01625-2
117. N. Cui, B. Luan, H. K. Liu, H. J. Zhao, and S. X. Dou, "Characteristics of magnesium-based hydrogen-storage alloy electrodes," *J Power Sources*, vol. 55, no. 2, pp. 263–267, Jun. 1995. doi: 10.1016/0378–7753(95)02195-M
118. H. Liu, J. Zhang, P. Sun, C. Zhou, Y. Liu, and Z. Z. Fang, "An overview of TiFe alloys for hydrogen storage: structure, processes, properties, and applications," *J Energy Storage*, vol. 68, p. 107772, Sep. 2023. doi: 10.1016/j.est.2023.107772
119. S. Semboshi, N. Masahashi, and S. Hanada, "Effect of composition on hydrogen absorbing properties in binary $TiMn_2$ based alloys," *J Alloys Compd*, vol. 352, no. 1–2, pp. 210–217, Mar. 2003. doi: 10.1016/S0925–8388(02)01125-8
120. T. Gamo, Y. Moriwaki, N. Yanagihara, T. Yamashita, and T. Iwaki, "Formation and properties of titanium-manganese alloy hydrides," *Int J Hydrogen Energy*, vol. 10, no. 1, pp. 39–47, 1985. doi: 10.1016/0360–3199(85)90134-X
121. H. Fujii, S. Munehiro, K. Fujii, and S. Orimo, "Effect of mechanical grinding under Ar and H_2 atmospheres on structural and hydriding properties in $LaNi_5$," *J Alloys Compd*, vol. 330–332, pp. 747–751, Jan. 2002. doi: 10.1016/S0925–8388(01)01508-0
122. P. Konik et al., "Structure and hydrogenation features of mechanically activated $LaNi_5$-type alloys," *Int J Hydrogen Energy*, vol. 46, no. 25, pp. 13638–13646, Apr. 2021. doi: 10.1016/j.ijhydene.2020.07.163
123. N. A. Ali and M. Ismail, "Modification of $NaAlH_4$ properties using catalysts for solid-state hydrogen storage: a review," *Int J Hydrogen Energy*, vol. 46, no. 1, pp. 766–782, Jan. 2021. doi: 10.1016/j.ijhydene.2020.10.011
124. B. Bogdanović and M. Schwickardi, "Ti-doped alkali metal aluminium hydrides as potential novel reversible hydrogen storage materials," *J Alloys Compd*, vol. 253–254, pp. 1–9, May 1997. doi: 10.1016/S0925–8388(96)03049-6
125. O. Friedrichs et al., "Direct synthesis of $Li[BH_4]$ and $Li[BD_4]$ from the elements," *Acta Mater*, vol. 56, no. 5, pp. 949–954, Mar. 2008. doi: 10.1016/j.actamat.2007.10.055
126. J. Huot, D. B. Ravnsbæk, J. Zhang, F. Cuevas, M. Latroche, and T. R. Jensen, "Mechanochemical synthesis of hydrogen storage materials," *Prog Mater Sci*, vol. 58, no. 1, pp. 30–75, Jan. 2013. doi: 10.1016/j.pmatsci.2012.07.001
127. A. S. Aricò, P. Bruce, B. Scrosati, J.-M. Tarascon, and W. van Schalkwijk, "Nanostructured materials for advanced energy conversion and storage devices," *Nat Mater*, vol. 4, no. 5, pp. 366–377, May 2005. doi: 10.1038/nmat1368

128. Z. Chen, K. O. Kirlikovali, K. B. Idrees, M. C. Wasson, and O. K. Farha, "Porous materials for hydrogen storage," *Chem*, vol. 8, no. 3, pp. 693–716, Mar. 2022. doi: 10.1016/j.chempr.2022.01.012
129. D. A. Gómez-Gualdrón et al., "Understanding volumetric and gravimetric hydrogen adsorption trade-off in metal–organic frameworks," *ACS Appl Mater Interfaces*, vol. 9, no. 39, pp. 33419–33428, Oct. 2017. doi: 10.1021/acsami.7b01190
130. Z. Chen et al., "Reticular access to highly porous acs-MOFs with rigid trigonal prismatic linkers for water sorption," *J Am Chem Soc*, vol. 141, no. 7, pp. 2900–2905, Feb. 2019. doi: 10.1021/jacs.8b13710
131. A. J. Cairns et al., "Gaining insights on the H_2 –sorbent interactions: robust soc-MOF platform as a case study," *Chem Mater*, vol. 28, no. 20, pp. 7353–7361, Oct. 2016. doi: 10.1021/acs.chemmater.6b02817
132. Y. Liu et al., "Assembly of Metal–Organic Frameworks (MOFs) based on indium-trimer building blocks: a porous MOF with soc topology and high hydrogen storage," *Angew Chem Int Ed*, vol. 46, no. 18, pp. 3278–3283, Apr. 2007. doi: 10.1002/anie.200604306
133. M. T. Kapelewski et al., "M_2 (*m*-dobdc) (M=Mg, Mn, Fe, Co, Ni) metal–organic frameworks exhibiting increased charge density and enhanced H_2 binding at the open metal sites," *J Am Chem Soc*, vol. 136, no. 34, pp. 12119–12129, Aug. 2014. doi: 10.1021/ja506230r
134. W. Zhou, H. Wu, and T. Yildirim, "Enhanced H_2 adsorption in isostructural metal–organic frameworks with open metal sites: strong dependence of the binding strength on metal ions," *J Am Chem Soc*, vol. 130, no. 46, pp. 15268–15269, Nov. 2008. doi: 10.1021/ja807023q
135. D. E. Jaramillo et al., "Ambient-temperature hydrogen storage via Vanadium(II)-dihydrogen complexation in a metal–organic framework," *J Am Chem Soc*, vol. 143, no. 16, pp. 6248–6256, Apr. 2021. doi: 10.1021/jacs.1c01883
136. D. Denysenko, M. Grzywa, J. Jelic, K. Reuter, and D. Volkmer, "Scorpionate-type coordination in MFU-4 *l* metal–organic frameworks: small-molecule binding and activation upon the thermally activated formation of open metal sites," *Angew Chem Int Ed*, vol. 53, no. 23, pp. 5832–5836, Jun. 2014. doi: 10.1002/anie.201310004
137. F. Li, J. Zhao, B. Johansson, and L. Sun, "Improving hydrogen storage properties of covalent organic frameworks by substitutional doping," *Int J Hydrogen Energy*, vol. 35, no. 1, pp. 266–271, Jan. 2010. doi: 10.1016/j.ijhydene.2009.10.061
138. E. Klontzas, E. Tylianakis, and G. E. Froudakis, "Hydrogen storage in lithium-functionalized 3-D covalent-organic framework materials," *J Phys Chem* C, vol. 113, no. 50, pp. 21253–21257, Dec. 2009. doi: 10.1021/jp907241y
139. D. Cao, J. Lan, W. Wang, and B. Smit, "Lithium-doped 3D covalent organic frameworks: high-capacity hydrogen storage materials," *Angew Chem Int Ed*, vol. 48, no. 26, pp. 4730–4733, Jun. 2009. doi: 10.1002/anie.200900960
140. K. E. Lamb, M. D. Dolan, and D. F. Kennedy, "Ammonia for hydrogen storage; a review of catalytic ammonia decomposition and hydrogen separation and purification," *Int J Hydrogen Energy*, vol. 44, no. 7, pp. 3580–3593, Feb. 2019. doi: 10.1016/j.ijhydene.2018.12.024
141. V. Negro, M. Noussan, and D. Chiaramonti, "The potential role of ammonia for hydrogen storage and transport: a critical review of challenges and opportunities," *Energies*, vol. 16, no. 17, p. 6192, Aug. 2023. doi: 10.3390/en16176192
142. C. Smith, A. K. Hill, and L. Torrente-Murciano, "Current and future role of Haber–Bosch ammonia in a carbon-free energy landscape," *Energy Environ Sci*, vol. 13, no. 2, pp. 331–344, 2020. doi: 10.1039/C9EE02873K
143. N. Pernicone, "Wustite as a new precursor of industrial ammonia synthesis catalysts," *Appl Catal A Gen*, vol. 251, no. 1, pp. 121–129, Sep. 2003. doi: 10.1016/S0926-860X(03)00313-2

144. F. Rosowski, A. Hornung, O. Hinrichsen, D. Herein, M. Muhler, and G. Ertl, "Ruthenium catalysts for ammonia synthesis at high pressures: preparation, characterization, and power-law kinetics," *Appl Catal A Gen*, vol. 151, no. 2, pp. 443–460, Apr. 1997. doi: 10.1016/S0926-860X(96)00304-3
145. H. Liu, *Ammonia Synthesis Catalysts*. World Scientific/Chemical Industry Press, China, 2013. doi: 10.1142/8199
146. H. S. Ahmed, Z. Yahya, W. Ali khan, and A. Faraz, "Sustainable pathways to ammonia: a comprehensive review of green production approaches," *Clean Energy*, vol. 8, no. 2, pp. 60–72, Apr. 2024. doi: 10.1093/ce/zkae002
147. T. Wu, W. Fan, Y. Zhang, and F. Zhang, "Electrochemical synthesis of ammonia: progress and challenges," *Mater Today Phys*, vol. 16, p. 100310, Jan. 2021. doi: 10.1016/j.mtphys.2020.100310
148. W. Xu et al., "Nanoporous palladium hydride for electrocatalytic N_2 reduction under ambient conditions," *Angew Chem Int Ed*, vol. 59, no. 9, pp. 3511–3516, Feb. 2020. doi: 10.1002/anie.201914335
149. W. Tong, B. Huang, P. Wang, L. Li, Q. Shao, and X. Huang, "Crystal-phase-engineered PdCu electrocatalyst for enhanced ammonia synthesis," *Angew Chem Int Ed*, vol. 59, no. 7, pp. 2649–2653, Feb. 2020. doi: 10.1002/anie.201913122
150. C. Yang, B. Huang, S. Bai, Y. Feng, Q. Shao, and X. Huang, "A generalized surface chalcogenation strategy for boosting the electrochemical N_2 fixation of metal nanocrystals," *Adv Mater*, vol. 32, no. 24, p. 2001267, Jun. 2020. doi: 10.1002/ADMA.202001267
151. G. N. Schrauzer and T. D. Guth, "Photolysis of water and photoreduction of nitrogen on titanium dioxide," *J Am Chem Soc*, vol. 99, no. 22, pp. 7189–7193, Oct. 1977. doi: 10.1021/ja00464a015
152. C. Liu et al., "Knowledge-driven design and lab-based evaluation of B-doped TiO_2 photocatalysts for ammonia synthesis," *Adv Energy Mater*, vol. 13, no. 8, Feb. 2023. doi: 10.1002/aenm.202204126
153. Y. Shi et al., "Engineering photocatalytic ammonia synthesis," *Chem Soc Rev*, vol. 52, no. 20, pp. 6938–6956, 2023. doi: 10.1039/D2CS00797E
154. J. Di et al., "Defect-tailoring mediated electron–hole separation in single-unit-cell Bi_3O_4Br nanosheets for boosting photocatalytic hydrogen evolution and nitrogen fixation," *Adv Mater*, vol. 31, no. 28, Jul. 2019. doi: 10.1002/adma.201807576
155. S. Wang et al., "Light-switchable oxygen vacancies in ultrafine Bi_5O_7Br nanotubes for boosting solar-driven nitrogen fixation in pure water," *Adv Mater*, vol. 29, no. 31, Aug. 2017. doi: 10.1002/adma.201701774
156. T. He et al., "Porphyrin-based covalent organic frameworks anchoring Au single atoms for photocatalytic nitrogen fixation," *J Am Chem Soc*, vol. 145, no. 11, pp. 6057–6066, Mar. 2023. doi: 10.1021/JACS.2C10233/ASSET/IMAGES/LARGE/JA2C10233_0007.JPEG
157. K. An et al., "Nitrogenase-inspired bimetallic metal organic frameworks for visible-light-driven nitrogen fixation," *Appl Catal B*, vol. 292, p. 120167, Sep. 2021. doi: 10.1016/J.APCATB.2021.120167
158. H. Shen et al., "Electrochemical ammonia synthesis: mechanistic understanding and catalyst design," *Chem*, vol. 7, no. 7, pp. 1708–1754, Jul. 2021. doi: 10.1016/j.chempr.2021.01.009
159. J. Zhao and Z. Chen, "Single Mo atom supported on defective boron nitride monolayer as an efficient electrocatalyst for nitrogen fixation: a computational study," *J Am Chem Soc*, vol. 139, no. 36, pp. 12480–12487, Sep. 2017. doi: 10.1021/jacs.7b05213
160. C. Ling, X. Bai, Y. Ouyang, A. Du, and J. Wang, "Single molybdenum atom anchored on N-doped carbon as a promising electrocatalyst for nitrogen reduction into ammonia at ambient conditions," *J Phys Chem* C, vol. 122, no. 29, pp. 16842–16847, Jul. 2018. doi: 10.1021/acs.jpcc.8b05257

161. Y. Yao et al., "Electrochemical synthesis of ammonia from nitrogen under mild conditions: current status and challenges," *Electrochem Energy Rev*, vol. 3, no. 2, pp. 239–270, Jun. 2020. doi: 10.1007/s41918-019-00061-3
162. M. Pinzón, R. García-Carpintero, A. R. de la Osa, A. Romero, D. Abad-Correa, and P. Sánchez, "Ammonia as a hydrogen carrier: an energy approach," *Energy Convers Manag*, vol. 321, Dec. 2024. doi: 10.1016/j.enconman.2024.118998
163. L. Zhou, L. Zhong, Z. Liu, and H. Wei, "Toward highly-efficient combustion of ammonia–hydrogen engine: prechamber turbulent jet ignition," *Fuel*, vol. 352, p. 129009, Nov. 2023. doi: 10.1016/j.fuel.2023.129009
164. K. Trangwachirachai, K. Rouwenhorst, L. Lefferts, and J. A. Faria Albanese, "Recent progress on ammonia cracking technologies for scalable hydrogen production," *Curr Opin Green Sustain Chem*, vol. 49, p. 100945, Oct. 2024. doi: 10.1016/j.cogsc.2024.100945
165. N. M. Adli, H. Zhang, S. Mukherjee, and G. Wu, "Review – ammonia oxidation electrocatalysis for hydrogen generation and fuel cells," *J Electrochem Soc*, vol. 165, no. 15, pp. J3130–J3147, Sep. 2018. doi: 10.1149/2.0191815jes
166. J. E. Lee, J. Lee, H. Jeong, Y.-K. Park, and B.-S. Kim, "Catalytic ammonia decomposition to produce hydrogen: a mini-review," *Chem Eng J*, vol. 475, p. 146108, Nov. 2023. doi: 10.1016/j.cej.2023.146108
167. S. Ristig et al., "Ammonia decomposition in the process chain for a renewable hydrogen supply," *Chem Ing Tech*, vol. 94, no. 10, pp. 1413–1425, Oct. 2022. doi: 10.1002/cite.202200003
168. S. Sun et al., "Ammonia as hydrogen carrier: advances in ammonia decomposition catalysts for promising hydrogen production," *Renew Sustain Energy Rev*, vol. 169, p. 112918, Nov. 2022. doi: 10.1016/j.rser.2022.112918
169. F. Schüth, R. Palkovits, R. Schlögl, and D. S. Su, "Ammonia as a possible element in an energy infrastructure: catalysts for ammonia decomposition," *Energy Environ. Sci.*, vol. 5, no. 4, pp. 6278–6289, 2012. doi: 10.1039/C2EE02865D
170. I. Lucentini, X. Garcia, X. Vendrell, and J. Llorca, "Review of the decomposition of ammonia to generate hydrogen," *Ind Eng Chem Res*, vol. 60, no. 51, pp. 18560–18611, Dec. 2021. doi: 10.1021/acs.iecr.1c00843
171. F. R. García-García, A. Guerrero-Ruiz, and I. Rodríguez-Ramos, "Role of B5-type sites in Ru catalysts used for the NH_3 decomposition reaction," *Top Catal*, vol. 52, no. 6–7, pp. 758–764, Jun. 2009. doi: 10.1007/s11244-009-9203-7
172. C. Chen et al., "Ru-based catalysts for ammonia decomposition: a mini-review," *Energy Fuels*, vol. 35, no. 15, pp. 11693–11706, Aug. 2021. doi: 10.1021/acs.energyfuels.1c01261
173. W. Zheng, J. Zhang, H. Xu, and W. Li, "NH_3 decomposition kinetics on supported Ru clusters: morphology and particle size effect," *Catal Lett*, vol. 119, no. 3–4, pp. 311–318, Nov. 2007. doi: 10.1007/s10562-007-9237-z
174. A. M. Karim et al., "Correlating particle size and shape of supported Ru/γ-Al_2 O_3 catalysts with NH_3 decomposition activity," *J Am Chem Soc*, vol. 131, no. 34, pp. 12230–12239, Sep. 2009. doi: 10.1021/ja902587k
175. S.-F. Yin, Q.-H. Zhang, B.-Q. Xu, W.-X. Zhu, C.-F. Ng, and C.-T. Au, "Investigation on the catalysis of COx-free hydrogen generation from ammonia," *J Catal*, vol. 224, no. 2, pp. 384–396, Jun. 2004. doi: 10.1016/j.jcat.2004.03.008
176. P. Yu et al., "Effects of alkaline earth metal amides on Ru in catalytic ammonia decomposition," *J Phys Chem* C, vol. 120, no. 5, pp. 2822–2828, Feb. 2016. doi: 10.1021/acs.jpcc.5b11768
177. J. Zhang, H. Xu, Q. Ge, and W. Li, "Highly efficient Ru/MgO catalysts for NH_3 decomposition: synthesis, characterization and promoter effect," *Catal Commun*, vol. 7, no. 3, pp. 148–152, Mar. 2006. doi: 10.1016/j.catcom.2005.10.002

178. X. Ju et al., "Mesoporous Ru/MgO prepared by a deposition-precipitation method as highly active catalyst for producing COx-free hydrogen from ammonia decomposition," *Appl Catal B*, vol. 211, pp. 167–175, Aug. 2017. doi: 10.1016/j.apcatb.2017.04.043
179. Y. Li et al., "Core–shell structured microcapsular-like Ru@SiO_2 reactor for efficient generation of COx-free hydrogen through ammonia decomposition," *ChemComm*, vol. 46, no. 29, p. 5298, 2010. doi: 10.1039/c0cc00430h
180. D. Varisli and E. E. Elverisli, "Synthesizing hydrogen from ammonia over Ru incorporated SiO_2 type nanocomposite catalysts," *Int J Hydrogen Energy*, vol. 39, no. 20, pp. 10399–10408, Jul. 2014. doi: 10.1016/j.ijhydene.2014.04.207
181. Z. Wang, Y. Qu, X. Shen, and Z. Cai, "Ruthenium catalyst supported on Ba modified ZrO_2 for ammonia decomposition to COx-free hydrogen," *Int J Hydrogen Energy*, vol. 44, no. 14, pp. 7300–7307, Mar. 2019. doi: 10.1016/j.ijhydene.2019.01.235
182. A. Klerke, S. K. Klitgaard, and R. Fehrmann, "Catalytic ammonia decomposition over ruthenium nanoparticles supported on nano-titanates," *Catal Lett*, vol. 130, no. 3–4, pp. 541–546, Jul. 2009. doi: 10.1007/s10562-009-9964-4
183. Z. Hu, J. Mahin, S. Datta, T. E. Bell, and L. Torrente-Murciano, "Ru-based catalysts for H_2 production from ammonia: effect of 1D support," *Top Catal*, vol. 62, no. 17–20, pp. 1169–1177, Nov. 2019. doi: 10.1007/s11244-018-1058-3
184. S. J. Wang, S. F. Yin, L. Li, B. Q. Xu, C. F. Ng, and C. T. Au, "Investigation on modification of Ru/CNTs catalyst for the generation of COx-free hydrogen from ammonia," *Appl Catal B*, vol. 52, no. 4, pp. 287–299, Oct. 2004. doi: 10.1016/j.apcatb.2004.05.002
185. H. Muroyama, C. Saburi, T. Matsui, and K. Eguchi, "Ammonia decomposition over Ni/La_2O_3 catalyst for on-site generation of hydrogen," *Appl Catal A Gen*, vol. 443–444, pp. 119–124, Nov. 2012. doi: 10.1016/j.apcata.2012.07.031
186. I. Lucentini, A. Casanovas, and J. Llorca, "Catalytic ammonia decomposition for hydrogen production on Ni, Ru and Ni Ru supported on CeO_2," *Int J Hydrogen Energy*, vol. 44, no. 25, pp. 12693–12707, May 2019. doi: 10.1016/j.ijhydene.2019.01.154
187. Q.-F. Deng, H. Zhang, X.-X. Hou, T.-Z. Ren, and Z.-Y. Yuan, "High-surface-area $Ce_{0.8}Zr_{0.2}O_2$ solid solutions supported Ni catalysts for ammonia decomposition to hydrogen," *Int J Hydrogen Energy*, vol. 37, no. 21, pp. 15901–15907, Nov. 2012. doi: 10.1016/j.ijhydene.2012.08.069
188. K. Okura, K. Miyazaki, H. Muroyama, T. Matsui, and K. Eguchi, "Ammonia decomposition over Ni catalysts supported on perovskite-type oxides for the on-site generation of hydrogen," *RSC Adv*, vol. 8, no. 56, pp. 32102–32110, 2018. doi: 10.1039/C8RA06100A
189. K. Okura, T. Okanishi, H. Muroyama, T. Matsui, and K. Eguchi, "Ammonia decomposition over nickel catalysts supported on rare-Earth oxides for the on-site generation of hydrogen," *ChemCatChem*, vol. 8, no. 18, pp. 2988–2995, Sep. 2016. doi: 10.1002/cctc.201600610
190. "Metal membrane for hydrogen separation." https://www.csiro.au/en/research/environmental-impacts/fuels/hydrogen/Membrane
191. Y. Yang et al., "A rigorous electrochemical ammonia electrolysis protocol with *in operando* quantitative analysis," *J Mater Chem A Mater*, vol. 9, no. 19, pp. 11571–11579, 2021. doi: 10.1039/D1TA00363A
192. E. Latvytė, X. Zhu, L. Wu, R. Lan, P. Vale, and J. E. Graves, "A low-temperature ammonia electrolyser for wastewater treatment and hydrogen production," *Int J Hydrogen Energy*, vol. 52, pp. 265–282, Jan. 2024. doi: 10.1016/j.ijhydene.2023.05.076
193. A. AlZaabi, F. AlMarzooqi, and D. Choi, "Ammonia electro-catalysis for hydrogen production: mechanisms, materials, and scalability," *Int J Hydrogen Energy*, vol. 94, pp. 23–52, Dec. 2024. doi: 10.1016/j.ijhydene.2024.11.082
194. A. AlZaabi, F. AlMarzooqi, and D. Choi, "Ammonia electro-catalysis for hydrogen production: mechanisms, materials, and scalability," *Int J Hydrogen Energy*, vol. 94, pp. 23–52, Dec. 2024. doi: 10.1016/j.ijhydene.2024.11.082

195. F. J. Vidal-Iglesias, J. Solla-Gullón, J. M. Feliu, H. Baltruschat, and A. Aldaz, "DEMS study of ammonia oxidation on platinum basal planes," *J Electroanal Chem*, vol. 588, no. 2, pp. 331–338, Mar. 2006. doi: 10.1016/j.jelechem.2006.01.009
196. F. J. Vidal-Iglesias, J. Solla-Gullón, J. M. Pérez, and A. Aldaz, "Evidence by SERS of azide anion participation in ammonia electrooxidation in alkaline medium on nano-structured Pt electrodes," *Electrochem Commun*, vol. 8, no. 1, pp. 102–106, Jan. 2006. doi: 10.1016/j.elecom.2005.11.002
197. D. Skachkov, C. Venkateswara Rao, and Y. Ishikawa, "Combined first-principles molecular dynamics/density functional theory study of ammonia electrooxidation on Pt(100) electrode," *J Phys Chem* C, vol. 117, no. 48, pp. 25451–25466, Dec. 2013. doi: 10.1021/jp4048874
198. L. A. Diaz and G. G. Botte, "Mathematical modeling of ammonia electrooxidation kinetics in a Polycrystalline Pt rotating disk electrode," *Electrochim Acta*, vol. 179, pp. 519–528, Oct. 2015. doi: 10.1016/j.electacta.2014.12.162
199. J. F. E. Gootzen, A. H. Wonders, W. Visscher, R. A. van Santen, and J. A. R. van Veen, "A DEMS and cyclic voltammetry study of NH_3 oxidation on platinized platinum," *Electrochim Acta*, vol. 43, no. 12–13, pp. 1851–1861, May 1998. doi: 10.1016/S0013–4686(97)00285-5
200. A. C. A. de Vooys, M. T. M. Koper, R. A. van Santen, and J. A. R. van Veen, "The role of adsorbates in the electrochemical oxidation of ammonia on noble and transition metal electrodes," *J Electroanal Chem*, vol. 506, no. 2, pp. 127–137, Jun. 2001. doi: 10.1016/S0022–0728(01)00491-0
201. L. ZHOU and Y. CHENG, "Catalytic electrolysis of ammonia on platinum in alkaline solution for hydrogen generation," *Int J Hydrogen Energy*, vol. 33, no. 21, pp. 5897–5904, Nov. 2008. doi: 10.1016/j.ijhydene.2008.08.009
202. J. Gwak, M. Choun, and J. Lee, "Alkaline ammonia electrolysis on electrodeposited platinum for controllable hydrogen production," *ChemSusChem*, vol. 9, no. 4, pp. 403–408, Feb. 2016. doi: 10.1002/cssc.201501046
203. J. Liu, W. Hu, C. Zhong, and Y. F. Cheng, "Surfactant-free electrochemical synthesis of hierarchical platinum particle electrocatalysts for oxidation of ammonia," *J Power Sources*, vol. 223, pp. 165–174, Feb. 2013. doi: 10.1016/j.jpowsour.2012.09.074
204. Z.-F. Li, Y. Wang, and G. G. Botte, "Revisiting the electrochemical oxidation of ammonia on carbon-supported metal nanoparticle catalysts," *Electrochim Acta*, vol. 228, pp. 351–360, Feb. 2017. doi: 10.1016/j.electacta.2017.01.020
205. F. J. Vidal-Iglesias, N. García-Aráez, V. Montiel, J. M. Feliu, and A. Aldaz, "Selective electrocatalysis of ammonia oxidation on Pt(100) sites in alkaline medium," *Electrochem Commun*, vol. 5, no. 1, pp. 22–26, Jan. 2003. doi: 10.1016/S1388–2481(02)00521-0
206. F. J. Vidal-Iglesias, J. Solla-Gullón, V. Montiel, J. M. Feliu, and A. Aldaz, "Ammonia selective oxidation on Pt(100) sites in an alkaline medium," *J Phys Chem* B, vol. 109, no. 26, pp. 12914–12919, Jul. 2005. doi: 10.1021/jp051269d
207. R. A. Martínez-Rodríguez, F. J. Vidal-Iglesias, J. Solla-Gullón, C. R. Cabrera, and J. M. Feliu, "Synthesis of Pt nanoparticles in water-in-oil microemulsion: effect of HCl on their surface structure," *J Am Chem Soc*, vol. 136, no. 4, pp. 1280–1283, Jan. 2014. doi: 10.1021/ja411939d
208. E. Moran et al., "Ammonia oxidation on electrodeposited Pt–Ir alloys," *J Solid State Electrochem*, vol. 12, no. 5, pp. 583–589, May 2008. doi: 10.1007/s10008-007-0407-0
209. Y. Zhou et al., "High mass and specific activity for ammonia electro-oxidation through optimization of dispersion degree and particle size of Pt-Ir nanoparticles over N-doped reductive graphene oxide," *ChemistrySelect*, vol. 3, no. 12, pp. 3433–3443, Mar. 2018. doi: 10.1002/slct.201800168
210. S. A. Lee, M. G. Lee, and H. W. Jang, "Catalysts for electrochemical ammonia oxidation: trend, challenge, and promise," *Sci China Mater*, vol. 65, no. 12, pp. 3334–3352, Dec. 2022. doi: 10.1007/s40843-022-2111–2

211. A. J. Medford et al., "From the Sabatier principle to a predictive theory of transition-metal heterogeneous catalysis," *J Catal*, vol. 328, pp. 36–42, Aug. 2015. doi: 10.1016/j.jcat.2014.12.033
212. A. Yang, J. Wang, K. Su, W. Lei, X. Qiu, and Y. Tang, "Modulating hydroxyl-rich interfaces on Nickel–Copper double hydroxide nanotyres to pre-activate alkaline ammonia oxidation reactivity," *Chem Eur J*, vol. 27, no. 15, pp. 4869–4875, Mar. 2021. doi: 10.1002/chem.202004904
213. W. Xu et al., "Directly growing hierarchical nickel-copper hydroxide nanowires on carbon fibre cloth for efficient electrooxidation of ammonia," *Appl Catal B*, vol. 218, pp. 470–479, Dec. 2017. doi: 10.1016/j.apcatb.2017.07.005
214. X. Jiang et al., "Identification of the role of Cu site in Ni-Cu hydroxide for robust and high selective electrochemical ammonia oxidation to nitrite," *Electrochim Acta*, vol. 345, p. 136157, Jun. 2020. doi: 10.1016/j.electacta.2020.136157
215. J. Moszczyńska, X. Liu, and M. Wiśniewski, "Green hydrogen production through ammonia decomposition using non-thermal plasma," *Int J Mol Sci*, vol. 24, no. 18, p. 14397, Sep. 2023. doi: 10.3390/ijms241814397
216. Y. Gorbanev, I. Fedirchyk, and A. Bogaerts, "Plasma catalysis in ammonia production and decomposition: use it, or lose it?," Curr Opin Green Sustain Chem, vol. 47, p. 100916, Jun. 2024. doi: 10.1016/j.cogsc.2024.100916
217. Z. Wang et al., "Plasma-promoted ammonia decomposition over supported ruthenium catalysts for CO_x -free H_2 production," *ChemSusChem*, vol. 16, no. 24, Dec. 2023. doi: 10.1002/cssc.202202370
218. X. Yu et al., "Plasma-catalytic ammonia decomposition for carbon-free hydrogen production using low pressure-synthesized Mo_2N catalyst," *Plasma Chem Plasma Process*, vol. 43, no. 1, pp. 183–197, Jan. 2023. doi: 10.1007/s11090-022-10282-y
219. Q. F. Lin et al., "Instantaneous hydrogen production from ammonia by non-thermal arc plasma combining with catalyst," *Energy Rep*, vol. 7, pp. 4064–4070, Nov. 2021. doi: 10.1016/j.egyr.2021.06.087

6 Barriers in Industrial Implementation of Next-Gen Carbon Capture Materials

Abhishek Bhatnagar and Divya Bhatnagar

6.1 INTRODUCTION

Large-scale production and consumption of goods and services is necessary for economic growth to cater to the changing needs of society. However, this production leads to high levels of CO_2 emissions that has placed immense pressure on industries worldwide to develop and implement effective carbon reduction strategies. Among the most promising solutions for mitigating these emissions are carbon capture, utilization, and storage technologies [1]. Over the past decade, significant progress has been made in the development of next-generation carbon capture materials (NG-CCMs), including metal–organic frameworks (MOFs), advanced sorbents, and other innovative materials. These materials offer several advantages, such as high selectivity for CO_2 capture, low regeneration energy requirements, and tunable properties that make them highly adaptable to various industrial applications. However, despite their performance in laboratory and pilot-scale studies, NG-CCMs face significant barriers that impede their large-scale deployment [2].

This chapter explores the industrial implementation of NG-CCMs with respect to the barriers that hinder their integration with real-world industrial processes. By identifying and analyzing the key barriers—technical, economic, logistical, and, policy and regulatory—the chapter aims to provide a comprehensive understanding of the complexities involved in transitioning from laboratory-scale innovations to full-scale industrial adoption of NG-CCMs.

The chapter is organized into seven sections as described below:

- Section 2**: Types of Carbon Capture Materials:** This section details out different NG-CCMs currently under development, such as MOFs, advanced sorbents, and other emerging materials. It provides their properties, advantages, and limitations in the context of industrial use.
- Section 3**: Carbon Capture Technologies:** This provides various technologies employed for carbon capture. Each technology is described along with its characteristics for carbon capture.

DOI: 10.1201/9781003582304-8

- Section 4: **Barriers to Industrial Implementation:** This section identifies the key barriers to the large-scale implementation of NG-CCMs. These include technical challenges (such as material stability and regeneration issues), economic hurdles (such as high production costs and return on investment uncertainty), logistical obstacles (including supply chain complexity and deployment infrastructure), and policy and regulatory challenges (such as lack of standards and inconsistent policy incentives).
- Section 5: **Interdisciplinary Solutions:** This section proposes potential solutions to the barriers discussed in Section 4, emphasizing the importance of interdisciplinary approaches. Solutions range from collaborative research and development (R&D) and advanced modeling to policy advocacy and standardization.
- Section 6: **Case Studies:** This section provides practical insights into how industries are grappling with the implementation of NG-CCMs. Four case studies from cement, paint, automotive, and logistics industries help understand the barriers and potential solutions.
- Section 7: **Conclusion:** The final section summarizes the chapter and offers insights into the future of NG-CCMs in industrial applications.

By examining these key areas, this chapter aims to provide a holistic understanding of the barriers to the industrial implementation of NG-CCMs, while offering interdisciplinary solutions that could accelerate their commercialization and scale-up.

6.2 TYPES OF CARBON CAPTURE MATERIALS

The selection of the right materials for carbon capture is a critical step to ensure efficiency, scalability, and economic viability of carbon capture technologies. The current available NG-CCMs offer a range of properties that can be tailored for specific industrial applications to provide significant potential for CO_2 capture, storage, and utilization [2]. These materials include MOFs, advanced sorbents, ionic liquids, carbon nanotubes (CNTs), and graphene-based materials [3] as given in Sections 2.1–2.5.

6.2.1 Metal–Organic Frameworks (MOFs)

MOFs are a class of highly porous crystalline materials made from metal nodes coordinated to organic linkers, forming a three-dimensional network. The exceptional surface area, tunable porosity, and versatile chemical functionality make MOFs one of the most promising candidates for CO_2 capture. However, high porosity and chemical structure of MOFs can make them vulnerable to degradation under harsh industrial conditions, including high temperature, humidity, and exposure to impurities like SO_2 and NO_x. Their scaling up is costly due to complex synthesis process. Further, many MOFs require high temperatures for effective regeneration, which also increases the energy costs [2]. MOFs are best suited for applications where their high selectivity toward CO_2 and tunable properties can be maximized, such as in high-purity CO_2 streams from industries like natural gas processing or post-combustion capture in power plants [1].

6.2.2 Advanced Sorbents

Advanced sorbents capture CO_2 either by physical adsorption or chemical absorption. They are widely used in both pre-combustion and post-combustion carbon capture processes and are among the most promising materials for industrial CO_2 capture due to their relatively low cost and high efficiency [2]. Different types of sorbents are as follows [1]:

- **Amine-functionalized sorbents**: These materials use amine groups to chemically bind CO_2, offering high CO_2 uptake and selectivity. They are widely studied for both post-combustion capture and direct air capture (DAC) technologies.
- **Zeolites**: These are crystalline aluminosilicates used in CO_2 capture due to their high selectivity for CO_2, stability, and tunable pore sizes. They are often utilized for pre-combustion CO_2 capture, particularly in natural gas and hydrogen production.

Sorbents, particularly zeolites, tend to be less expensive to produce compared to MOFs. Many sorbents, such as carbon-based materials, can be easily regenerated by pressure or temperature swing, which helps reduce operational costs. However, some sorbents, particularly amine-functionalized sorbents, can undergo degradation under high temperatures or exposure to water vapor, limiting their long-term viability. While sorbents can be regenerated, the energy requirements for the regeneration process (such as temperature swing or pressure swing adsorption) can significantly reduce their efficiency. In some cases, the adsorption capacity of sorbents may be lower than that of MOFs, making them less suitable for high-concentration CO_2 streams unless engineered to improve performance.

6.2.3 Ionic Liquids

Ionic liquids (ILs) are salts that are liquid at room temperature and are composed of large, complex ions. They are highly polar and exhibit low volatility, making them excellent candidates for CO_2 capture. They have negligible vapor pressure, which reduces losses during CO_2 capture. The ionic composition of ILs can be adjusted to enhance CO_2 absorption that offer a high degree of customization for different industrial processes. ILs can dissolve CO_2 at high concentrations, making them suitable for large-scale CO_2 capture. However, ILs tend to have high viscosity, which sometimes lead to slower absorption rates and higher energy costs during CO_2 capture. The production of high-purity ILs is expensive, and their commercialization for large-scale applications needs significant cost reductions [4]. Some ILs are toxic or may have adverse environmental impacts if not disposed of properly, which could restrict their widespread adoption.

6.2.4 Carbon Nanotubes

CNTs are cylindrical structures made from graphene sheets. They possess unique mechanical, thermal, and electrical properties and have shown potential for CO_2 adsorption due to their large surface area and high porosity. Their electrical properties make CNTs useful for applications where the material needs to be integrated into electronic systems, such as carbon capture in sensors or CO_2 detectors. While CNTs show great potential for CO_2 capture, the mass production of high-quality CNTs remains a challenge, primarily due to expensive production methods. To enhance CO_2 selectivity, CNTs often need to be functionalized with various groups, which is a complex and costly process [5]. There are concerns about the environmental impact of CNTs, particularly regarding their potential toxicity if released into the atmosphere or environment.

6.2.5 Graphene and Graphene Oxide

Graphene is a single layer of carbon atoms arranged in a hexagonal lattice. Due to its extraordinary surface area, mechanical strength, and conductivity, graphene and its derivatives (such as graphene oxide) have emerged as promising candidates for CO_2 capture [6]. Graphene and graphene oxide possess extremely high surface areas, which can enhance their CO_2 uptake capabilities. Graphene-based materials are lightweight, strong, and flexible, making them suitable for integration into a variety of industrial applications. Functionalization of graphene oxide allows for the modification of its properties to improve its selectivity and capture efficiency for CO_2. Despite the benefits, high-quality graphene is expensive to produce, and current manufacturing techniques are not yet optimized for large-scale production.

6.3 CARBON CAPTURE TECHNOLOGIES

Carbon capture technologies are essential for reduction of greenhouse gas emissions, especially in industries that generate significant amount of CO_2. These technologies can be broadly categorized into three primary areas: post-combustion capture, pre-combustion capture, DAC and ocean fertilization and algae-based methods. Each method has its peculiar characteristics, and the choice of technology often depends on factors such as the type of emission source, the cost of implementation, and energy consumption. These methods are covered in detail in Sections 6.3.1–6.3.4.

6.3.1 Post-Combustion Capture

Post-combustion capture involves capturing CO_2 from the flue gases emitted by power plants or industrial facilities after fossil fuels are burned. This is one of the most widely applied methods for carbon capture and offers the benefit of retrofitting existing facilities. In this method, CO_2 is separated from the flue gases using solvents, adsorbents, or membranes. The most commonly used method is solvent-based absorption, where a liquid absorbent (usually amine-based) chemically binds to CO_2, removing it from the gas stream [7]. The advantages of this process include the following:

- **Retrofit potential**: The technology can be applied to existing power plants and industrial facilities, making it an attractive option for reducing emissions from legacy infrastructure.

- **Proven technology**: Post-combustion capture is already being implemented in several pilot projects and commercial facilities, demonstrating its feasibility.
- **Versatility**: This method is suitable for a wide range of industries, including power generation, cement production, and steel manufacturing.

The regeneration of solvents used in absorption-based post-combustion capture processes requires significant amounts of energy, often derived from fossil fuels, which can undermine the overall environmental benefit. Over time, solvents can degrade and form by-products, reducing efficiency and increasing operational costs. For large-scale applications, significant infrastructure, including absorber columns and solvent regeneration units, is required, which increases costs and complexity.

6.3.2 Pre-Combustion Capture

Pre-combustion capture is generally applied in gasification processes, where fossil fuels (such as coal or natural gas) are converted into syngas (a mixture of hydrogen and carbon monoxide) before combustion. CO_2 is captured from the syngas stream before the fuel is burnt, typically through processes like chemical looping or membrane separation. This generally occurs in gasification or integrated gasification combined cycle plants. The process involves removing CO_2 from the syngas before combustion, which can be achieved via physical or chemical solvents or by using membrane separation techniques [7]. The advantages of this process include the following:

- **Higher CO_2 concentration**: It works with syngas, which has a higher concentration of CO_2, making the capture process more efficient.
- **Improved efficiency**: Since CO_2 is removed before combustion, power plants using pre-combustion capture tend to have higher overall efficiency.
- **Flexibility**: This technology can be applied to a variety of fuel sources, including coal, natural gas, and biomass, making it adaptable for different energy systems.

This technology requires significant investment in gasification and syngas clean-up infrastructure, which increases initial capital costs. The complexity of gasification processes and syngas treatment poses operational challenges, particularly in scaling up to commercial levels. While pre-combustion processes tend to be more energy-efficient than post-combustion capture, they still require a considerable amount of energy to separate CO_2 from the syngas.

6.3.3 Direct Air Capture (DAC)

DAC focuses on capturing CO_2 directly from the ambient air rather than from concentrated industrial emissions. DAC technologies are typically used with either chemical or physical sorbents that selectively capture CO_2 from atmospheric concentrations, which are significantly lower than those found in flue gases. These use either liquid or solid sorbents for CO_2 capture. In the liquid-based approach, a chemical sorbent (such as amine solution) absorbs CO_2 from the air, which is then regenerated by heating [8]. In the solid-based approach, a solid sorbent, typically a porous material like a MOF or an advanced sorbent, adsorbs CO_2, and is regenerated by heat or pressure swing. The advantages of this technology include the following:

- **Scalability**: It has potential to scale globally since it can be deployed anywhere, even in areas far from emission sources. This makes it an attractive option for removing CO_2 from the atmosphere on a global scale.
- **Flexibility**: It is not limited to specific industries and can be used to capture CO_2 from various locations, including remote regions and even the oceans.
- **Negative emissions**: It can be used as negative emissions technology which enables the removal of historical CO_2 emissions from the atmosphere and help mitigate climate change.

The major drawback of DAC is the high energy required to capture CO_2 from the low-concentration ambient air. This results in high operational costs and energy use, particularly in regions where energy prices are high. Large-scale deployment may require a significant land area for the installation of equipment, which could compete with land needed for food production or conservation. Its cost is high compared to other carbon capture methods, limiting its commercial viability.

6.3.4 Ocean Fertilization and Algae-Based Technologies

Ocean fertilization refers to the addition of nutrients such as iron to the ocean to stimulate the growth of algae that absorb CO_2 during photosynthesis. Algae-based CO_2 capture technologies aim to leverage biological processes to remove CO_2 from the atmosphere. Algae-based carbon capture can occur both in open ocean systems and in controlled systems such as artificial algae bioreactors that can both efficiently use lighting and minimize energy costs [9]. The captured CO_2 is then either stored in biomass or converted into biofuels, such as algae-derived bioethanol or biodiesel. The advantages of this method include the following:

- **Sustainability**: These technologies harness natural biological processes and have the potential to be sustainable if managed correctly.
- **Scalable**: Microalgae cultivation is done on an industrial scale in an open pond with a shallow depth and large area for growth of microcultures.
- **Co-benefits**: Algae cultivation also has the potential to provide biofuels, making it an attractive option for both carbon capture and renewable energy generation.

6.4 BARRIERS TO INDUSTRIAL IMPLEMENTATION OF NG-CCMS

While the potential of NG-CCMs to mitigate CO_2 emissions is immense, the path to their industrial implementation comes with several barriers and challenges. These barriers span multiple domains, including technical, economic, logistical, and policy regulatory factors as given in Sections 6.4.1–6.4.4 and given in Figure 6.1. Understanding and addressing these challenges are critical for accelerating the adoption of NG-CCMs on a global scale.

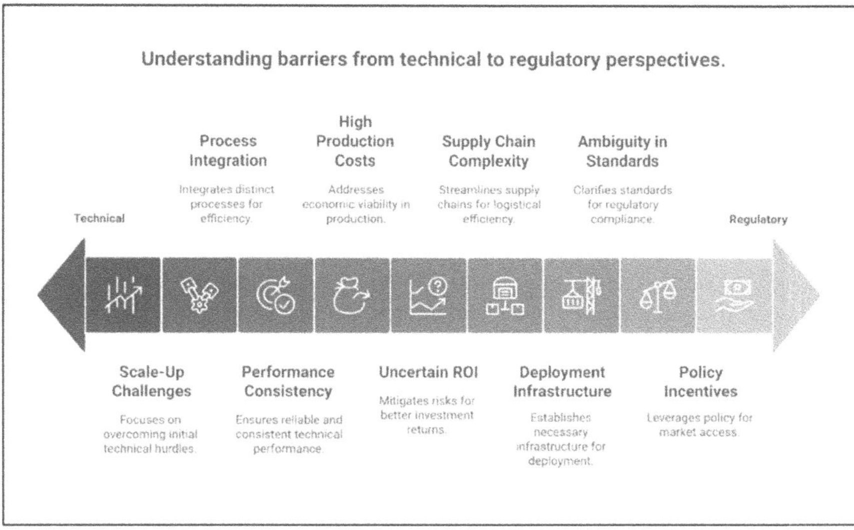

FIGURE 6.1 Barriers to adoption of NG-CCMs.

6.4.1 Technical Barriers

The technical barriers to industrial deployment of NG-CCMs are very predominant, given the complexity of scaling laboratory innovations to industrial levels. While NG-CCMs exhibit outstanding performance in controlled laboratory conditions, translating these results to real-world applications poses several hurdles. The technical barriers are further elaborated in the following subsections.

6.4.1.1 Scale-Up Challenges

The ability to scale NG-CCMs from laboratory-scale demonstrations to full-scale industrial operations is one of the most pressing challenges. These challenges are as follows:

- **Material stability**: NG-CCMs often degrade or lose efficiency under extreme conditions such as high temperature, humidity, or exposure to impurities commonly found in industrial emissions. For instance, MOFs, while highly effective in CO_2 capture, are prone to degradation in the presence of water vapor, necessitating constant optimization of material formulations.
- **Chemical deactivation**: Many NG-CCMs, particularly sorbents and solvents, undergo chemical deactivation over time. This degradation can lead to decreased efficiency and higher operational costs due to the need for regular replenishment or regeneration.
- **Long-term durability**: The long-term durability of NG-CCMs under cyclic operation conditions (e.g., repeated cycles of CO_2 capture and release) is another major concern. Performance degradation over time, often linked to fouling or structural changes in the material, can severely impact the economics of the technology.

6.4.1.2 Process Integration

The integration of NG-CCMs into existing industrial processes often requires significant changes to established operations, which can be both technically challenging and costly.

- **Compatibility with industrial gases**: Industrial flue gases vary significantly in composition, including the presence of impurities such as sulfur compounds, particulate matter, and other gases (e.g., nitrogen oxides), which can compromise the efficiency of NG-CCMs. These materials must be robust enough to operate in diverse environments and resist poisoning from these contaminants.
- **System retrofits**: Many industries would need to retrofit existing infrastructure to accommodate NG-CCMs. For instance, integrating carbon capture systems into cement or steel plants often requires redesigning flue gas handling systems, increasing space requirements, and adding additional equipment such as absorption towers or membrane filtration units.

6.4.1.3 Performance Consistency

Achieving consistent performance across different industrial settings and conditions is a significant challenge.

- **Gas composition variability**: Variations in the chemical composition of flue gases (e.g., CO_2 concentrations, temperature, pressure) pose challenges in ensuring NG-CCMs operate efficiently across different industrial sectors.
- **Regeneration energy costs**: Many NG-CCMs require significant energy to regenerate the sorbent or absorbent material after CO_2 capture. This regeneration process often involves heating or pressurization, which leads to substantial energy consumption, reducing overall system efficiency. The need for additional energy input for regeneration can undermine the benefits of net carbon reduction.

6.4.2 Economic Barriers

Economic factors play a critical role in determining the feasibility of NG-CCMs for industrial deployment. The high cost of production, the uncertain financial returns, and the lack of clear incentives present formidable challenges to widespread adoption.

6.4.2.1 High Production Costs

The manufacturing cost of NG-CCMs is currently one of the most significant barriers to industrial adoption.

- **Raw material costs**: Many NG-CCMs, such as MOFs and advanced sorbents, require rare or expensive raw materials, including metals and specialized chemical compounds. The high cost of these materials contributes significantly to the overall cost of production.

- **Complex synthesis processes**: The synthesis of advanced materials often involves complex chemical processes that are not optimized for large-scale production. For example, the synthesis of MOFs typically requires the use of solvents and reagents that are not readily available in bulk, increasing both material and labor costs.
- **Scalability of manufacturing**: Currently, most NG-CCMs are produced in small quantities for research purposes, and scaling up production to meet industrial demands remains a challenge. Finding cost-effective manufacturing methods that maintain material quality at scale is critical for making NG-CCMs commercially viable.

6.4.2.2 Uncertain Return on Investment (ROI)

The financial viability of NG-CCMs remains uncertain due to several factors given below:

- **High upfront capital investment**: The initial cost of implementing NG-CCMs in industrial facilities, including the purchase of materials, retrofitting existing infrastructure, and installing carbon capture systems, can be prohibitively high. Many companies are hesitant to invest in technologies with uncertain returns, particularly when the payback period is long.
- **Fluctuating carbon credit prices**: The economic incentive for implementing carbon capture technologies often depends on the availability of carbon credits or carbon pricing mechanisms. However, the price of carbon credits can be volatile, leading to financial uncertainty for industries investing in NG-CCMs.
- **Unpredictable policy support**: The financial viability of NG-CCMs also depends on government policies, such as subsidies for carbon capture research, tax credits, and carbon pricing schemes. The inconsistency or lack of clear policy frameworks across regions can dissuade investment in NG-CCMs.

6.4.3 Logistical Barriers

The logistics surrounding the deployment of NG-CCMs introduce additional complexities, from material sourcing to the transportation and storage of sensitive materials. Addressing these logistical challenges is crucial for ensuring the scalability and effectiveness of carbon capture solutions.

6.4.3.1 Supply Chain Complexity

- **Raw material availability**: The extraction of raw materials required for NG-CCMs, such as rare earth elements and metals (e.g., copper, zinc, aluminum), often depends on complex and geographically concentrated supply chains. Disruptions in these supply chains, whether due to geopolitical factors or environmental considerations, could hinder the availability of materials necessary for NG-CCM production.

- **Material handling and transport**: Many NG-CCMs are sensitive to environmental factors such as temperature and humidity. Ensuring that these materials are transported and stored under optimal conditions can add complexity and cost to the overall supply chain. The handling and shipping of materials with specialized requirements could increase both direct and indirect costs.

6.4.3.2 Deployment Infrastructure

- **Retrofitting existing facilities**: Industries such as cement, steel, and power generation often lack the infrastructure necessary for the installation of NG-CCM systems. Retrofitting existing plants requires significant investment in both physical infrastructure and engineering design. Moreover, the retrofitting process can cause operational disruptions and delays in the short term.
- **Lack of standardized deployment systems**: The deployment of NG-CCMs varies significantly between industries, and the lack of standardized equipment and deployment procedures can lead to inefficiencies. Each industrial facility may require bespoke systems for carbon capture, adding to the complexity and costs involved.
- **Storage and disposal of captured CO_2**: Once CO_2 is captured, it must be either stored or utilized. The lack of sufficient infrastructure for safe and effective CO_2 storage (such as geological storage sites) or for utilization (in the production of valuable chemicals or fuels) can create a bottleneck in the carbon capture process.

6.4.4 Regulatory and Policy Barriers

Regulatory frameworks and government policies play a significant role in the deployment of NG-CCMs. However, inconsistencies and ambiguities in regulations can hinder the large-scale adoption of carbon capture technologies.

6.4.4.1 Ambiguity in Standards

- **Lack of certification protocols**: The absence of universally accepted standards for evaluating and certifying NG-CCMs complicates their adoption. Industry stakeholders need clear guidelines for assessing the efficiency, durability, and environmental impact of NG-CCMs. Without such standards, companies may be hesitant to invest in technologies that have not been fully vetted.
- **Variability in regulatory approaches**: Different countries and regions have varying approaches to carbon capture regulation. Inconsistent regulatory frameworks across jurisdictions can create barriers to the international adoption of NG-CCMs, as companies face uncertainty in complying with multiple and sometimes conflicting regulations.

6.4.4.2 Policy Incentives and Market Mechanisms

- **Inconsistent carbon pricing**: The financial incentive for implementing NG-CCMs often hinges on carbon pricing mechanisms, which can vary greatly between regions. In regions where carbon pricing is either absent or poorly implemented, industries may have little motivation to adopt carbon capture technologies.

- **Limited government support**: While some governments provide financial support for R&D in carbon capture, few offer substantial subsidies or incentives for the deployment of NG-CCMs at the industrial level. The lack of consistent policy support hinders the financial attractiveness of carbon capture solutions.

6.5 INTERDISCIPLINARY SOLUTIONS FOR THE IMPLEMENTATION OF NG-CCMS

Overcoming the barriers to the large-scale deployment of NG-CCMs requires a multifaceted, interdisciplinary approach. Given the complex nature of the challenges—spanning technical, economic, logistical, and regulatory domains—it is essential to integrate expertise from diverse fields such as materials science, mechanical engineering, operations research, and policy development. This section explores several potential solutions that combine cutting-edge research, technological innovation, and collaborative efforts between academia, industry, and government.

6.5.1 Collaborative Research and Development (R&D)

Collaborative R&D initiatives between academic institutions, private companies, and governmental organizations can provide a significant boost to overcoming the technical and economic barriers to NG-CCM deployment. By fostering partnerships across disciplines and sectors, stakeholders can share resources, expertise, and knowledge, enabling the development of innovative solutions that might be out of reach for individual entities. Such collaboration can focus on improving the material properties of NG-CCMs, making them more robust under industrial conditions, and optimizing their performance for large-scale applications.

For example, universities and research institutions can work closely with industry partners to develop cost-effective synthesis methods for NG-CCMs, ensuring that production processes are scalable and economical. This may involve optimizing the use of readily available raw materials, exploring novel synthesis routes, or developing new processing techniques to reduce production costs. Furthermore, academic researchers can collaborate with mechanical engineers and operations researchers to better understand the long-term durability of NG-CCMs under real-world operational conditions. This would involve modeling their performance across different industrial settings and identifying material compositions that minimize degradation and maximize efficiency over time.

6.5.2 Advanced Modeling and Simulation

Advanced modeling and simulation techniques, particularly those derived from the field of operations research, are invaluable tools for optimizing the deployment and performance of NG-CCMs. Operations researchers use mathematical modeling, optimization algorithms, and simulations to improve decision-making, streamline operations, and minimize costs. These tools can be applied to the design, scaling, and integration of NG-CCMs within industrial processes, enabling companies to better assess their feasibility before large-scale implementation.

One of the most promising applications of modeling is in the optimization of supply chains for NG-CCMs. Given the complexity of sourcing raw materials, manufacturing the materials, transporting them, and deploying them at industrial facilities, operations research techniques can help create more efficient supply chains. Optimization models can evaluate multiple variables, such as transportation costs, lead times, and material availability, to identify the most cost-effective routes and methods for distributing NG-CCMs. Furthermore, these models can account for potential disruptions in the supply chain, such as changes in raw material prices or geopolitical events that affect material availability.

Moreover, advanced modeling can help evaluate the life-cycle costs and environmental impact of NG-CCMs. By simulating the entire carbon capture process, including material production, CO_2 capture, regeneration, and disposal, operations researchers can calculate the total energy consumption, carbon footprint, and financial outlay associated with NG-CCMs. This comprehensive analysis can guide industries in making informed investment decisions and identifying areas where cost reductions and performance improvements are most needed.

Simulations can also be used to predict the performance of NG-CCMs under various industrial conditions, such as fluctuating gas compositions, temperatures, and pressures. This can aid in the design of more resilient materials and ensure that the NG-CCMs can consistently deliver high performance across a range of operational scenarios.

6.5.3 Policy Advocacy and Standardization

For NG-CCMs to be widely adopted in industry, it is crucial that governments and policymakers develop clear, supportive policies that incentivize investment in carbon capture technologies. Policy advocacy plays a vital role in bridging the gap between scientific advancements and industrial implementation. One of the key policy measures needed is the establishment of standardized testing and certification protocols for NG-CCMs. These standards would allow companies to evaluate the performance and safety of materials in a consistent, reliable manner. Without such standards, industries face uncertainty regarding the technical performance and durability of NG-CCMs, making them reluctant to invest in these technologies.

In addition, governments can play a significant role in mitigating the financial risks associated with adopting NG-CCMs. For example, tax incentives, grants, or direct subsidies can offset the initial capital investment required to implement carbon capture systems. In particular, financial support for early adopters—who are often the first to integrate innovative technologies—can accelerate the transition to NG-CCM adoption. Furthermore, government policies that support R&D for NG-CCMs can help lower the cost of these technologies by funding long-term innovation and facilitating knowledge transfer between academic researchers and industrial practitioners.

Carbon pricing mechanisms, such as carbon taxes or cap-and-trade systems, also provide an economic incentive for companies to adopt carbon capture technologies. However, to be effective, these systems must be consistent, predictable, and appropriately calibrated to reflect the true environmental cost of carbon emissions. Policymakers must work to create a stable and favorable economic environment for NG-CCMs by ensuring that carbon pricing schemes are well-structured and aligned with the global push to reduce greenhouse gas emissions.

Beyond economic incentives, governments can drive the adoption of NG-CCMs by implementing regulatory frameworks that require industries to adopt carbon capture technologies. For example, emission reduction mandates for heavy industries, such as cement, steel, and energy production, can create a strong regulatory push for companies to integrate NG-CCMs into their operations. Clear and enforceable standards would help industries navigate the complex regulatory landscape and ensure compliance with environmental goals.

6.5.4 Financial and Market Mechanisms

The financial challenges associated with NG-CCMs are significant, but they can be mitigated through the development of robust financial and market mechanisms. One promising solution is the creation of financing models that reduce the upfront capital costs of implementing NG-CCMs. These could include performance-based funding, where the financial support is tied to the successful demonstration of technology's efficacy at scale. Additionally, blended finance models—combining public and private investments—can be used to fund large-scale deployment projects, especially in industries that are traditionally risk-averse, such as cement or steel production.

Another important financial mechanism is the expansion of carbon credit markets. As NG-CCMs directly contribute to reducing CO_2 emissions, companies that adopt these technologies should be eligible to earn carbon credits that can be sold in global carbon markets. This would create an additional revenue stream for companies, improving the financial viability of NG-CCMs. To ensure the effective functioning of carbon credit markets, it is essential to establish transparent and verifiable emission reduction methodologies. This would provide companies with the confidence that their efforts to capture CO_2 are being accurately measured and rewarded.

Further financial incentives can be provided through long-term power purchase agreements for low-carbon energy sources or through contracts that guarantee the purchase of captured CO_2 for utilization in chemical production or enhanced oil recovery. This would provide certainty for investors and businesses regarding the future market demand for carbon capture technologies.

6.6 CASE STUDIES: INDUSTRY APPLICATIONS AND BARRIERS TO NG-CCM ADOPTION

The implementation of NG-CCMs across various industries offers valuable insights into the challenges and opportunities associated with large-scale adoption. In this section, we explore four industries—cement, paint, automotive, and logistics. These case studies illustrate how each sector faces unique barriers to implementation and the innovative strategies being developed to overcome these hurdles.

6.6.1 Cement Industry: Integrating MOFs for CO_2 Capture

The cement industry is a major contributor to global CO_2 emissions, due to the calcination of limestone during cement production. This makes it a critical target for carbon capture technologies. MOFs are considered promising due to their high surface area, adjustable porosity, and capacity to capture CO_2 at low concentrations typical of cement flue gases. However, their deployment faces significant technical, economic, and logistical barriers.

Technically, MOFs struggle with stability under the extreme heat and variable gas compositions found in cement kilns. Regenerating MOFs to maintain continuous capture also demands high energy input, reducing their practicality. Economically, MOFs are expensive to produce, largely due to costly metal precursors, and scaling production for industrial use is financially daunting. Uncertain returns from carbon pricing mechanisms further hinder investment.

Logistically, retrofitting existing cement plants with CO_2 capture systems involves complex redesigns of gas handling and heat integration systems, along with challenges related to CO_2 transport and storage infrastructure. Despite these barriers, collaborative research is underway to improve MOF performance at high temperatures and explore alternative materials with greater thermal stability. Policy support through carbon pricing is expected to enhance economic viability. In the long term, effective integration of carbon capture technologies like MOFs is seen as vital for the cement industry to meet sustainability and emissions reduction goals.

6.6.2 Paint Industry: Developing Advanced Sorbents for CO_2 Capture

The paint industry, though not a major direct source of CO_2 emissions, contributes to greenhouse gases through energy-intensive manufacturing and solvent use. As global demand for paints grows, the industry faces increasing pressure to reduce its environmental impact, including exploring carbon capture technologies. Advanced sorbents, which selectively capture CO_2 from low-concentration gas streams, show promise but face several hurdles.

Technical challenges include ensuring sorbents remain effective in the presence of volatile organic compounds and can perform under the temperature and humidity conditions typical of paint manufacturing. Economic barriers involve the high costs of developing and deploying these materials, especially given the low CO_2 concentrations in paint industry emissions. Logistical issues include space limitations and integration into existing facilities, as well as the need for CO_2 transport and storage infrastructure.

To overcome these obstacles, ongoing R&D is focusing on improving sorbent regeneration and cost-effectiveness, while also developing materials suitable for low-CO_2 environments. Policy support and technological innovation are expected to play key roles in enabling the adoption of carbon capture in the paint industry.

6.6.3 Automotive Industry: CO_2 Capture for Fleet Emissions

The automotive industry is under growing pressure to lower carbon emissions from both vehicles and manufacturing. While electrification gets much attention, carbon capture technologies—particularly NG-CCMs—are being explored to reduce emissions from production plants and vehicle fuel combustion. However, NG-CCMs face technical challenges such as variable exhaust gas composition and high energy use, economic barriers like high capital costs and uncertain returns, and logistical issues involving plant retrofitting and CO_2 handling. To overcome these, research and collaboration are essential to develop efficient, scalable solutions. Carbon capture may become a key strategy in making the automotive sector more sustainable.

6.6.4 Logistics Industry: Enhancing Carbon Capture for Transportation Networks

The logistics industry, particularly freight transport via trucks, ships, and planes, is a major contributor to global CO_2 emissions. While cleaner fuels and electrification are being adopted, carbon capture technologies—specifically NG-CCMs—could help further reduce emissions by capturing CO_2 directly from vehicle exhausts. However, these systems face several challenges. Technically, NG-CCMs must be lightweight and compact to suit vehicles without affecting performance and must handle variable exhaust conditions. Economically, high installation costs and uncertain returns, compounded by low carbon credit prices, make adoption difficult. Logistically, effective CO_2 collection and storage infrastructure is required, and retrofitting fleets remains complex. Looking ahead, innovations in material science, supportive policy frameworks, and integration with alternative fuels or electrification could enhance the viability of NG-CCMs in decarbonizing logistics.

6.7 CONCLUSION

The chapter provided an overview of several carbon capture materials and techniques. It also discussed the barriers in the large-scale industrial implementation of these carbon capture materials, followed by case studies for four different sectors. These case studies from the cement, paint, automotive, and logistics industries highlight the barriers to the widespread adoption of NG-CCMs. Each sector faces unique technical, economic, and logistical challenges, but solutions are emerging through interdisciplinary collaboration, technological innovation, and supportive policies. As these industries work to overcome these barriers, the potential for NG-CCMs to contribute to global carbon mitigation goals remains significant.

REFERENCES

[1] B. Dziejarski, J. Serafin, K. Andersson, and R. Krzyżyńska, "CO_2 capture materials: A review of current trends and future challenges," *Mater. Today Sustain.*, vol. 24, 2023. doi: 10.1016/j.mtsust.2023.100483

[2] A. Allangawi, Abdulrahman, Eman FH Alzaimoor, Haneen H. Shanaah, Hawraa A. Mohammed, Husain Saqer, Ahmed Abd El-Fattah, and Ayman H. Kamel, "Carbon capture materials in post-combustion: Adsorption and absorption-based processes," *J. Carbon Res.*, vol. 9, no. 1, p. 17, 2023.

[3] M. Ozkan and R. Custelcean, "The status and prospects of materials for carbon capture technologies," *MRS Bull.*, vol. 47, no. 4, pp. 390–394, 2022. doi: 10.1557/s43577-022-00364-9

[4] M. Hasib-ur-Rahman, M. Siaj, and F. Larachi, "Ionic liquids for CO_2 capture—Development and progress," *Chem. Eng. Process. Process Intensif.*, vol. 49, no. 4, pp. 313–322, 2010. https://doi.org/10.1016/j.cep.2010.03.008

[5] P. Jayakumari, B. Y. Swaminathan, E. N. Partheeban, "A review on characteristics studies on carbon nanotubes-based cement concrete," *Constr. Build. Mater.*, vol. 367, p. 130344, 2023. https://doi.org/10.1016/j.conbuildmat.2023.130344

[6] L. Nagarajan, P. Sampath Kumar, A. Arulraj, G. Senguttuvan, and K. Kumaraguru, "Solar assisted reduced graphene oxide as adsorbent for carbon dioxide and its kinetic studies," *Phys. E Low-Dimensional Syst. Nanostructures*, vol. 116, no. March 2019, p. 113739, 2020. doi: 10.1016/j.physe.2019.113739

[7] F. Raganati and P. Ammendola, "CO_2 Post-combustion capture: A critical review of current technologies and future directions," *Energy and Fuels*, vol. 38, no. 15, pp. 13858–13905, 2024. doi: 10.1021/acs.energyfuels.4c02513

[8] X. Zhu et al., "Recent advances in direct air capture by adsorption," *Chem. Soc. Rev.*, vol. 51, no. 15, pp. 6574–6651, 2022. doi: 10.1039/d1cs00970b

[9] T. Iglina, P. Iglin, and D. Pashchenko, "Industrial CO_2 capture by algae: A review and recent advances," *Sustainability*, vol. 14, no. 7, 2022. doi: 10.3390/su14073801

7 Case Study
Solar Energy Potential in Sub-Saharan Africa towards the Energy Transition

Yahaya D. Baba, Alhassan H. Umar, Aliyu M. Aliyu, and Salem Brini Ahmed

7.1 INTRODUCTION

Sub-Saharan Africa (SSA) is an area of enormous potential and opportunities, but it is confronted with intimidating challenges which are a major obstacle to its socio-economic development. The current information state power supply in this mighty region of Africa is underpowered, with a substantial proportion of the inhabitants lacking access to modern, clean, and affordable energy services. The problem is clearly noticeable in rural communities where the dependence on traditional biomass such as wood and charcoal for cooking and heating remains high. In SSA, the traditional energy supply not only contributes to deforestation and environmental degradation but also poses serious health risks due to indoor air pollution, with major negative consequences on women and children [1]. Inadequate access to electricity further limits economic activity, education, and health services, prolonging the cycle of poverty. Addressing this energy gap is not only about providing energy; it is fundamental to achieving the sustainable development goals (SDGs), promoting inclusive growth and improving the overall quality of life of millions across the continent.

Due to the pressing need to improve energy security and mitigate climate change, there is a continuous global push to transition to cleaner, more sustainable energy sources. Solar energy has become one of the most alternative renewable energy sources in this global transition, because of its general availability, falling prices, and low environmental impact. It plays a crucial part in mitigating greenhouse gas emissions, decarbonising energy systems, and creating a resilient energy future. Solar energy can efficiently offer the SSA a singular chance to forge ahead of conventional fossil fuel-based development pathways, providing a decentralised and sustainable answer to the long-term inadequate energy access. The region has attracted greater attention because of its potential to influence paradigm shifts in energy.

DOI: 10.1201/9781003582304-9

7.1.1 Objectives of the Chapter

The purpose of this chapter is to present a comprehensive review of the solar energy potential in SSA, the challenges to its widespread adoption, and the opportunities it offers for a sustainable energy transition. The main objectives are as follows:

- Examine the current state of regional energy infrastructure, the prevalent energy mix, and access to contemporary energy in SSA.
- Assess the geographical and climatic benefits that highlight enormous potential of the sub-Saharan region for solar energy and review the current solar energy initiatives.
- Determine and critically analyse the main obstacles to the development of solar energy, including those related to funding, technology, policy, regulations, society, culture, and the environment.
- Explore the numerous advantages that solar energy provides for the energy transition of the area, such as its capacity to generate adequate electricity, improve living conditions, and induce technological innovations.
- Study and identify areas for improvement in the institutional, legal, and policy frameworks currently controlling the growth of solar energy.
- Make recommendations for effective policy, build institutional and technical capacity, and encourage regional cooperation to speed up the energy transition.

This chapter intends to uncover strategic interventions as well as provide comprehensive analysis to realise the most effective approach towards sustainable solar energy for the SSA.

7.2 ENERGY ECOSYSTEM IN SUB-SAHARAN AFRICA

7.2.1 Current Energy Mix

The mix of traditional and modern energy sources in SSA is intricate and often unbalanced. Even though the region has a lot of potential for renewable energy, traditional biomass, mostly wood, charcoal, and agricultural waste, dominates its energy consumption. This reliance on biomass makes up a large proportion of the primary energy supply, especially for domestic heating and cooking, especially in rural areas [1]. Unfortunately, over-reliance on traditional biomass has serious consequences for public health, environmental deterioration (deforestation), and socio-economic development.

When it comes to contemporary energy, fossil fuels and hydropower make up most of the electricity mix in SSA. In many nations, coal- and oil-fired power plants make significant contributions to the generation of grid-based electricity, though their proportion varies by region [2]. However, hydropower is the main low-carbon source and accounts for almost 29% of the electricity mix. In many SSA nations, hydropower is the main source of electricity [3]. Even with its substantial contribution, hydropower in Africa still has a lot of unrealised potential. Despite their quick rise, other renewable energy sources like wind and solar still make up a small portion of the total energy mix, except for conventional biomass [4]. Several East African nations, most notably Kenya, also use geothermal energy as part of their energy mix.

In SSA, the unequal distribution of infrastructure and energy resources also affects the energy mix. Although certain nations possess substantial fossil fuel reserves, others are highly dependent on imports, resulting in energy insecurity and susceptibility to worldwide price swings. To meet the region's increasing energy demand and attain universal energy access, it will be difficult to move away from its heavy reliance on fossil fuels and traditional biomass and towards a more diversified and sustainable energy mix that makes use of its wealth of renewable resources, especially solar.

7.2.2 Accessibility of Modern Energy

Access to contemporary energy services, especially clean cooking solutions and electricity, is still a major problem throughout SSA. Despite some advancements, the region still lags well behind global averages, which has serious ramifications for both economic growth and human development. SSA is thought to have had 600 million people without access to electricity in 2022, which is the largest percentage of people worldwide without electricity [5]. In rural areas, where electrification rates are much lower than in urban centres, this deficit is especially noticeable. Access may be restricted even in urban areas because of unregulated tariffs and the high cost of electricity derived from fossil fuels [6].

Access to clean cooking presents an even greater challenge. There are still about 700 million households in SSA that cook using traditional biomass fires [7]. This dependence on polluting and ineffective cooking techniques contributes to environmental deterioration and serious health problems, such as respiratory disorders. More than 93% of people in many nations, especially those living in rural areas, do not have access to clean cooking fuels and technologies [8]. The region has witnessed a persistent rise in the number of people using traditional cooking methods, despite initiatives to promote cleaner alternatives, underscoring the magnitude of the problem [9].

Wide-ranging effects result from not having access to contemporary energy. Since kids find it difficult to learn in dimly lit environments, it restricts educational opportunities. Healthcare facilities frequently lack dependable electricity, which affects both the standard of care and the capacity to stock necessary medications. Without reliable electricity, businesses cannot grow or function effectively, which limits economic activity. The SDGs, especially SDG 7, which seeks to guarantee that everyone has access to modern, affordable, dependable, and sustainable energy, depend on closing this energy access gap. While some areas have seen progress, especially with the emergence of off-grid solar solutions, electrification and the adoption of clean cooking must pick up speed considerably to meet the region's rising demand and raise living standards.

7.2.3 Regional Energy Infrastructure

A sustainable energy transition and the realisation of SSA's full energy potential depend on the establishment of strong and interconnected regional energy infrastructure. Because national grids frequently function in isolation, the region's energy infrastructure has historically been dispersed, resulting in inefficiencies, exorbitant costs, and an unstable power supply. Several regional power pools have been formed throughout SSA to address these issues, with the goals of promoting cross-border electricity trade and improving energy security.

The Central African Power Pool (CAPP), the East African Power Pool (EAPP), the West African Power Pool (WAPP), and the Southern African Power Pool (SAPP) are among the SSA's current operational or developing power pools [10]. These power pools are intended to increase transmission infrastructure and generation capacity, allowing nations to share electricity and balance supply and demand throughout their networks. By leveraging the region's varied energy resources, these pools enable nations with excess generating capacity to export electricity to those with shortages, maximising resource use and minimising the need for expensive national investments in redundant capacity.

These power pools are strategically important, but operationalising and growing them continues to present major obstacles. Within and between nations, the inadequate and deteriorating transmission infrastructure is a major obstacle. Upgrades to existing transmission lines and new transmission line investments are vital, but they frequently encounter difficult financing arrangements, drawn-out negotiations, and a lack of laws that encourage private sector involvement [11]. This lack of transmission infrastructure makes it more difficult to evacuate power from new sources of generation to demand centres, especially large-scale renewable energy projects situated in remote areas. Additionally, disparities in market structures, technical standards, and regulatory frameworks among nations may make it more difficult for these power pools to integrate and trade energy seamlessly.

A regional energy infrastructure that is well-integrated, however, has enormous potential advantages. Through grid stability improvements, power generation optimisation, and the elimination of costly peak-load plants, it can result in significant cost savings. Diversifying supply sources and lowering dependency on individual energy pathways are two more ways that regional integration improves energy security. Because regional power pools enable the balancing of supply and demand over a wider geographic area, they provide a mechanism to manage variability for renewable energy, especially intermittent sources like solar. A truly interconnected and resilient energy system in SSA will be realised by accelerating investment in regional transmission infrastructure and harmonising regulatory frameworks, opening the door to a more sustainable and prosperous future.

7.3 SOLAR ENERGY POTENTIAL IN SUB-SAHARAN AFRICA

7.3.1 Geographic and Climatic Advantages

SSA, with its unique geographical and climatic advantages, is uniquely placed to harness solar energy. It receives some of the highest levels of sunlight in the world, which makes it an ideal place to produce solar energy. Unlike most of the other continents, Africa has abundant sunshine throughout the year, with many regions getting more than 2500 hours of clear sunshine a year [12]. These steady and intensive suns translate into considerable potential for both photovoltaic (PV) and concentrated solar power (CSP) technologies.

SSA has far more solar radiation than it currently needs for electricity, which presents a wealth of opportunities for a range of uses, from large-scale power generation to home lighting [13]. The region's solar energy potential consistently outperforms that of many other regions worldwide, including those with substantially larger installed solar capacities, according to comparative analyses. SSA has a solid basis to move ahead of conventional energy development models and make the direct transition to a solar-powered future thanks to this innate natural endowment. Its appeal for solar energy investments is further increased by the region's stable and predictable solar resources, which in many places show little seasonal variation.

7.3.2 Solar Resource Availability

Due to its exceptional availability of solar resources, SSA is a prime location for the development of solar energy. The continent possesses about 60% of the world's best solar resources, but only 1% of the world's installed solar PV capacity is currently located there [14]. The region's immense unrealised potential is highlighted by this glaring disparity. Many parts of SSA receive 4–6 kWh m^{-2} day^{-1} of solar radiation on average, with some regions receiving even higher levels. The levels of solar radiation in many developed nations that have effectively implemented large-scale solar power are much lower than this. Germany, a pioneer in the use of solar energy, for example, receives an average of between 2.5 and 3.5 kWh m^{-2} per day in some areas.

In many parts of SSA, especially near the equator, the uniformity of the solar radiation throughout the year makes the area even more attractive. Unlike regions where there are strong seasonal variations, relatively stable resources simplify planning and reduce the need for extensive seasonal storage. That continued availability of solar energy makes it a reliable and predictable source of electricity for both grid and non-grid applications. The sheer amount of untapped energy means that solar alone could supply many times the continent's energy needs and provide a sustainable path to universal energy access and economic development.

To fully realise this potential and transform this plentiful natural resource into practical energy solutions, supportive policies and strategic investments are needed. Even though resources are abundant, the region's installed capacity is currently low, which emphasises the need for faster development and adoption of solar technologies. This involves reaching a variety of demographics and incorporating solar energy into the larger energy infrastructure by utilising both decentralised solutions and large-scale utility projects.

7.3.3 Existing Solar Energy Projects

SSA has seen a growing number of solar power projects ranging from small-scale decentralised solutions to large-scale utility-scale plants. Such projects are essential for tackling the region's energy deficit and for driving energy transformation.

7.3.3.1 Small-Scale Solar Solutions

Small-scale solar solutions, especially mini-grids and solar home systems (SHS), have become revolutionary instruments for SSA's rural electrification. Those communities and households that are not connected to the national grid can receive electricity from these decentralised systems. SHS provides basic electricity for small electronics, phone charging, and lighting, and usually consists of a solar panel, battery, and outlets for appliances and lighting [15]. In contrast, mini-grids are larger systems that produce and supply electricity to a group of residences and commercial establishments in a specific area, frequently enabling productive uses [16].

- **Kenya**: In terms of implementing small-scale solar solutions, especially SHS, Kenya has led the way. Numerous private companies and social enterprises are implementing SHS throughout rural areas, demonstrating the country's thriving off-grid solar market. Pay-as-you-go (PAYG) systems and other innovative financing models have empowered low-income households to access solar energy [17]. Additionally, solar-powered mini-grids have grown in popularity in Kenya, where rural residents are receiving power from private companies [18].
- **Tanzania**: Using mini-grids to electrify rural areas, Tanzania has also made notable progress. In Ketumbeine, for example, ENGIE Energy Access installed a 16 kWp mini-grid unit that supplies electricity to 120 residences and businesses [19]. These projects show how decentralised solar solutions can effectively close the gap in energy access.

7.3.3.2 Large-Scale Solar Projects

Large-scale solar projects, especially utility-scale solar farms, are diversifying energy mixes and improving energy security by increasingly adding to national grids throughout SSA. Both CSP and PV technologies are frequently used in these projects.

- **Noor Ouarzazate solar complex (Morocco):** Located in North Africa, this complex is a notable illustration of the continent's expansive solar ambition. The planned capacity of this complex exceeds 500 MW, making it one of the biggest CSP plants in the world. It makes it possible to generate electricity even after sunset by combining thermal storage and CSP [20]. Its success offers other African countries a model.
- **South Africa's REIPPPP:** With its Renewable Energy Independent Power Producer Procurement Programme (REIPPPP), South Africa has taken the lead in creating extensive renewable energy projects. Since its inception in 2011, this initiative has drawn substantial private funding, enabling the construction of more than 6 GW of utility-scale wind and solar projects [21]. Because of the REIPPPP's competitive bidding process, renewable energy is now more affordable and competitive.

These initiatives, no matter how big or small, are essential to supplying SSA's expanding energy needs, decreasing dependency on fossil fuels, and lowering greenhouse gas emissions.

7.4 CHALLENGES FACING SOLAR ENERGY DEVELOPMENT

Solar energy has enormous potential and is becoming increasingly popular in SSA, but there are several major obstacles that prevent its widespread development and use. It is imperative that these obstacles be removed if the area is to realise a sustainable energy transition and fully utilise its solar resources.

7.4.1 Financial Constraints

One major obstacle to the widespread use of solar energy in SSA is financial constraints. Whether they are small off-grid systems or large utility plants, the high upfront costs of solar projects frequently turn off potential developers and investors [22]. Solar energy has relatively low operating costs, but it can have high upfront costs for panels, inverters, batteries, and installation. This is especially difficult in an area where it can be difficult to find reasonably priced financing and where interest rates can be high.

Political unpredictability, exchange rate volatility, and lax regulations are among the perceived risks that make it challenging to secure both public and private investment for solar projects in SSA. The cost of financing solar projects rises even more because of investors' frequent demands for larger returns to offset these risks [23]. The financial feasibility of projects is greatly impacted by this "costly capital," which deters investors from investing in these areas in favour of others with more stable investment climates. Accessing financing is another difficulty for small-scale solar energy solutions like mini-grids and SHS. In rural areas, these solutions are essential for energy access; however, small businesses or individual households frequently lack the credit history or collateral needed to obtain loans. It will take creative financing schemes and de-risking techniques to close this gap and raise the substantial sum of money needed to expand solar energy in the area.

7.4.2 Technical Barriers

The extensive use and integration of solar energy in SSA is severely hampered by technical issues. The current energy infrastructure in many nations is frequently insufficient and not built to handle the erratic nature of renewable energy sources, such as solar. This poses problems for distribution, transmission, and grid stability.

The inadequate infrastructure for transmission and distribution is one of the biggest technical obstacles [24]. Numerous national grids in SSA are distinguished by their low capacity, ageing equipment, and restricted reach. This implies that, particularly in remote locations, the electricity produced by new solar power plants may not always be effectively transmitted to demand centres or distributed to end users. This problem is made worse by high technical losses that occur during transmission and distribution, resulting in energy waste and decreased system efficiency. Large-scale solar project integration into unreliable or underdeveloped grids may also result in voltage swings, grid instability, and reliability problems.

An additional significant technical obstacle is the erratic nature of solar power and the consequent requirement for reliable energy storage systems. Because solar power generation is reliant on sunlight availability, which varies throughout the day and is influenced by meteorological factors, it is inherently variable. Ensuring a steady and dependable power supply is difficult due to this intermittency, especially for systems that are connected to the grid. Utility-scale batteries are one type of energy storage solution that is crucial for grid stability and continuous power supply. But in most SSA nations, battery storage capacity is currently limited, and the high cost of these technologies continues to be a major barrier [25]. Without sufficient storage, solar energy's full potential to meet peak demand or supply baseload power cannot be achieved, making grid integration more difficult and expensive. For solar energy to be successfully integrated and used throughout the region, significant investment in smart grid technologies, grid modernisation, and the installation of cutting-edge energy storage systems are needed to overcome these technological obstacles.

7.4.3 Policy and Regulatory Gaps

Solar energy development in SSA is severely hampered by inconsistent energy policies and regulatory frameworks. In order to provide certainty and draw consistent investment in the renewable energy sector, many nations in the region lack long-term, stable, and well-defined energy policies [26]. A lack of transparency and frequent policy changes produce an unpredictable environment that discourages both domestic and foreign investors. Long-term planning, obtaining funding, and navigating the regulatory environment are all made more challenging for solar project developers by this policy instability.

Regulatory changes are urgently needed to promote more private sector participation in the solar energy industry. Current laws may be onerous, out-of-date, or inadequately adapted to the particulars of renewable energy projects. This can raise expenses and risks for developers by causing delays in project approvals, licensing, and grid connection [27]. Investment may also be discouraged by the difficulty solar projects may have securing revenue streams in the absence of strong and stable power purchase agreements (PPAs) and transparent tariff structures.

The scaling of solar energy projects depends heavily on government incentives and subsidies, but their application in SSA has frequently been uneven or inadequate. Although feed-in tariffs (FiTs), tax breaks, or import duty exemptions for solar equipment have been implemented in some nations, these policies are not consistently implemented or maintained. Clear renewable energy targets expedited permitting procedures, and carefully thought-out financial incentives are just a few examples of effective policy tools that are necessary to lower project costs, de-risk investments, and hasten the regional rollout of solar energy. The potential of solar energy in SSA will largely remain unrealised in the absence of a stable and supportive regulatory and policy environment.

7.4.4 Social and Cultural Factors

The widespread use of solar energy in SSA is hampered by social and cultural factors in addition to financial, policy, and infrastructure issues. Lack of knowledge and instruction about the features and advantages of solar energy technologies is a major obstacle [28]. Many communities might not be fully aware of how solar power operates, how much it can save in the long run, or how it can benefit the environment and human health, especially in rural areas. This lack of knowledge may cause people to be sceptical of and reluctant to adopt new energy solutions.

Deeply rooted cultural customs and social norms may also be the cause of resistance to implementing new energy technologies. For generations, people have relied on traditional energy sources like firewood and charcoal, which are frequently incorporated into daily activities and means of subsistence. It can be difficult to alter behaviour and perception when moving away from these well-known sources. A lack of confidence in more recent, decentralised solar solutions may lead to a preference for traditional grid electricity, even if it is unreliable or unavailable. Additionally, even a modest upfront cost for SHS can be a major deterrent for low-income households, who might put short-term demands ahead of long-term energy solutions.

In local communities, traditional energy sources play a versatile role, frequently acting as a source of fuel as well as an integral component of economic activity and cultural identity. Targeted awareness campaigns, community involvement initiatives, and educational programmes that sensitively and culturally highlight the concrete advantages of solar energy are necessary to overcome these social and cultural barriers. Fostering acceptance and speeding up adoption require establishing trust and proving the dependability and affordability of solar solutions locally [29].

7.4.5 Climate and Environmental Concerns

Despite its inherent cleanliness and ability to mitigate climate change, solar energy development and operation in SSA can be difficult due to its own environmental and climate concerns. To guarantee the sustainability of solar projects, these variables must be properly controlled.

The effect of dust and extremely high or low temperatures on solar panel performance is one major worry. The climates of many regions of SSA are arid or semi-arid, with high concentrations of sand and dust. When dust builds up on solar panels, it can drastically lower their efficiency and power output, requiring regular cleaning—which calls for water. In areas where water is scarce, this can pose a serious operational and environmental problem. Even though solar panels are made to run in the sun, extremely high temperatures, which are typical in many regions of SSA, can reduce their efficiency and affect the total amount of energy produced as well as the projects' financial feasibility [30].

Land use is an additional environmental factor, especially for large-scale solar farms. Large tracts of land are needed to develop utility-scale solar projects, which may put other land uses like grazing, agriculture, or conservation in competition. Inadequate planning and management of this could lead to habitat fragmentation, land displacement, and possible effects on biodiversity. Such projects' environmental impact assessments need to take these factors into account and investigate ways to use degraded lands or co-locate solar farms with other compatible land uses.

Additionally, one of the biggest environmental concerns is the amount of water used to clean solar panels in arid areas. Dust accumulation, as previously stated, makes routine cleaning necessary to preserve peak performance. This may further tax already limited water supplies, which could influence nearby communities and ecosystems. To lessen this difficulty, research and development into more effective cleaning techniques or waterless cleaning technologies is essential. An additional new environmental concern is how to properly dispose of solar panels at the end of their useful lives because doing so may cause dangerous materials to leak out. As the solar sector expands in the area, it will be crucial to create strong recycling regulations and infrastructure.

The implementation of sustainable practices throughout the life cycle of solar projects, careful planning, and the selection of appropriate technology are all necessary to address these climate and environmental concerns. SSA can guarantee the economic viability, environmental responsibility, and social equity of its solar energy development by proactively addressing these obstacles.

7.5 OPPORTUNITIES FOR SOLAR ENERGY IN ENERGY TRANSITION

SSA can benefit greatly from solar energy in both addressing its energy deficit and promoting a fair and sustainable energy transition. These opportunities cover social, economic, and environmental aspects.

7.5.1 Large-Scale Solar Farms

It is becoming more widely acknowledged that large-scale solar farms, especially utility-scale projects—are essential to SSA's energy transition. These projects are expensive and frequently use both CSP and PV technologies to produce electricity that is then widely distributed to national grids. Building such expansive facilities is essential to supplying the region's rapidly increasing energy needs, diversifying energy sources, and boosting energy security.

Some SSA countries are actively pursuing or have already commissioned large-scale solar power projects. South Africa, for example, has successfully facilitated the development of more than 6 GW of utility-scale solar and wind projects through its Renewable Energy Innovation and Production Partnership Programme (REIPPP) which has contributed significantly to its national grid. It has proven to be viable to attract private investments in large-scale renewable energy production by means of competitive tendering [31].

Other countries outside of South Africa are also making progress. For instance, Angola is now building several sizeable solar farms that should provide a sizeable amount of electricity and reach a sizeable population of households [32]. Another bold project that aims to build large-scale solar capacity across several Sahelian nations is the G5 Sahel Desert to Power project, which emphasises the possibility of regional cooperation in such endeavours [33]. In addition to producing electricity, these utility-scale projects support local economic growth, generate jobs during construction and operation, and promote skill transfer within the renewable energy industry [34].

These projects are a desirable alternative for increasing generation capacity because of the region's plentiful solar resources and the growing cost-competitiveness of large-scale solar PV. In line with international climate goals, they provide a clean substitute for power plants that rely on fossil fuels, greatly lowering greenhouse gas emissions. In order to provide steady, dependable, and clean energy to support industrialisation, urbanisation, and overall economic growth in SSA, large-scale solar farms will become even more crucial as grid infrastructure advances and regional power pools become stronger.

7.5.2 Solar as a Tool for Reducing Energy Poverty

The unmatched potential of solar energy to drastically lower energy poverty and increase access to electricity, especially in underserved rural areas, is one of the most revolutionary opportunities it offers SSA. Since a sizeable section of the population lacks access to dependable electricity, solar solutions provide a quick and decentralised deployment method for electrifying areas that are not expected to be connected to national grids anytime soon [35].

Mini-grids and SHS are becoming increasingly successful in providing electricity to these underserved communities. By giving businesses instant access to lighting, these solutions increase business productivity and enable children to study after dark, both of which enhance educational results. Reliable electricity also powers vital services like medical facilities, which allow for better lighting for patient care, refrigeration for medications and vaccines, and the use of medical equipment. Solar-powered irrigation pumps have the potential to transform farming methods in the agricultural industry, improving food security, lowering manual labour costs, and raising yields [36]. Additionally, having electricity makes communication and information access easier by enabling community centres to be powered by the internet and charging cell phones.

In addition to providing basic access, solar electrification has significant socioeconomic effects. By encouraging new businesses, allowing small businesses to work longer hours, and generating employment opportunities in the installation, upkeep, and distribution of solar products, it boosts local economies. Better indoor air pollution from conventional biomass fuels is also reduced by improved lighting and clean cooking options (when solar is combined with electric cooking), which improves health outcomes. Solar energy empowers communities and improves livelihoods by offering a dependable and reasonably priced energy source. It also directly supports several SDGs, such as those pertaining to gender equality, health, education, and the eradication of poverty. Because solar solutions are decentralised, development can start from scratch and be customised to meet the unique requirements and environments of various communities throughout the region [37]. Because solar energy directly reduces poverty, it is an essential instrument for sustainable and inclusive development in SSA.

7.5.3 Technological Innovations and Future Trends

The solar energy industry is known for its quick technological developments, which have great potential to hasten SSA's energy transition. Solar power is becoming more widely available, dependable, and affordable thanks to ongoing advancements in energy storage technologies, smart grid technologies, and solar panel efficiency.

Perovskite solar cells (PSCs): The advent of PSCs is one of the most exciting advancements in solar technology. In lab settings, these next-generation solar cells have demonstrated impressive efficiency gains, quickly catching up to and even surpassing the efficiency of conventional silicon-based solar cells [38]. Perovskites are also more adaptable and can be produced with simpler, less expensive methods, which could make them more economical and useful for a range of applications, such as flexible solar panels and building-integrated photovoltaics (BIPV). Their potential for high efficiency at reduced prices could make solar power in SSA even more competitive by drastically lowering its overall cost.

Floating solar (also known as floating photovoltaics, or FPV): These installations, which place solar panels on bodies of water like lakes, reservoirs, or even offshore, are becoming more and more popular all over the world. This technology has many benefits, especially in areas with expensive or limited land. FPV can prevent land-use conflicts, increase panel efficiency because of the cooling effect of water, and decrease evaporation from water bodies. With so many dams and reservoirs, floating solar offers SSA a practical way to increase solar capacity without displacing natural habitats or agricultural land [39].

Advanced energy storage solutions: Effective and reasonably priced energy storage solutions are required due to the intermittent nature of solar power. Research and development are still going on in several different battery chemistries and storage technologies, such as flow batteries, solid-state batteries, and thermal energy storage, even though lithium-ion batteries are currently the most common. Increased energy density, longer cycle life, lower costs, and enhanced safety are the goals of these developments. Better energy storage will be essential for SSA to increase grid stability, allow for greater solar power penetration, and provide a steady supply of electricity during times when solar irradiance is low [40].

Digitalisation and smart grids: One of the major trends of the future that will maximise the efficiency of solar energy systems is the combination of digitalisation and smart grid technologies. Smart grids track and control the flow of electricity in real time using sophisticated sensors, communication tools, and data analytics. Demand-side management, enhanced grid resilience, more accurate solar output forecasting, and the effective integration of dispersed renewable energy sources are all made possible by this. Digitalisation can improve solar operations, including the application of artificial intelligence and the Internet of Things [41].

For the SSA, incorporation of smart grids can help managing the transition from centralised energy systems to more decentralised energy systems. Thus, facilitating the integration of mini-grids and off-grid solutions with the main grid when feasible. These technological innovations, coupled with ongoing research and development, are continuously improving the performance, affordability, and versatility of solar energy. As these technologies mature and become more widely adopted, they will further enhance solar energy's role as a cornerstone of SSA's energy transition, enabling the region to build a more sustainable, resilient, and electrified future.

7.5.4 Material Innovations for Solar Energy

The recent innovations in material science are revolutionising the solar energy landscape, pushing the boundaries of efficiency, durability, and cost-effectiveness. These innovations have become particularly crucial for regions like SSA, where the unique environmental conditions and socio-economic contexts demand robust, affordable, and high-performing solar technologies. Beyond traditional silicon, a new generation of materials is emerging, promising to unlock unprecedented opportunities for solar energy deployment. Some of these materials are highlighted below.

PSCs are game changers for efficiency and cost: PSCs represent one of the most impactful breakthroughs in PV research in recent years [42]. The power conversion efficiency of these hybrid organic–inorganic compounds has increased dramatically, quickly catching up to and in some lab settings even surpassing that of traditional silicon solar cells [41]. What makes perovskites particularly attractive for SSA is their potential for cost efficiency in manufacturing. In contrast to silicon, which necessitates energy-intensive, high-temperature procedures, perovskite films can be produced at significantly lower temperatures through solution-based techniques like printing or coating. This significantly reduces production costs and energy consumption, making them a more affordable option for widespread deployment [43].

Beyond cost, perovskites also offer several other advantages. Their tuneable bandgap allows them to absorb a broader spectrum of sunlight, enhancing efficiency, especially in tandem configurations with silicon cells [44]. They are also lightweight and flexible, opening up possibilities for novel applications such as BIPV, flexible solar panels for portable devices, and even transparent solar cells for windows [45]. However, ongoing research is being done to address issues with long-term stability, especially in areas of SSA that have high temperatures and humidity levels, as well as the presence of lead in some formulations.

These stability concerns are being directly addressed by material innovations. Guanidinium salt addition to perovskite materials, for example, has demonstrated great promise in enhancing long-term stability and efficiency [43,45]. The incorporation of these salts modulates the crystal growth of the perovskite film, resulting in enlarged grain domains and reduced defect density, thereby facilitating more efficient charge extraction and enhancing overall device performance [46]. Alternative strategies currently under investigation include the development of lead-free perovskite compositions, the encapsulation of devices to mitigate environmental degradation, and the engineering of interfacial layers to enhance charge transport while suppressing recombination losses [47].

Harnessing energy with nanoscale precision using quantum dot solar cells (QDSCs): QDSCs are known to utilise semiconductor nanocrystals, typically 2–10 nm in size, to absorb sunlight and generate electricity [48]. They exhibit unique optoelectronic properties due to quantum mechanical effects, allowing for tuneable light absorption and emission characteristics simply by changing their size [49]. This property of spectral tunability enables QDSCs to be optimised for selective light harvesting across targeted wavelength ranges, thereby enhancing their efficiency in applications such as multi-junction solar cells and PV operation under low-light conditions. Their ability to be prepared in solution, similar to perovskites, further contributes to the potential for low-cost and scalable manufacturing [50].

For SSA, QDSCs present a promising pathway towards advancing sustainable energy solutions, offering the potential for highly efficient and mechanically flexible solar cells that can be adapted for diverse applications. Their tuneable properties and low-cost fabrication routes make them particularly suited for addressing energy access challenges in resource-constrained environments. Their performance in diffuse light conditions could also be advantageous in areas with frequent cloud cover or during early morning/late evening hours. Further research can be focused on improving the stability of quantum dots, developing non-toxic alternatives to cadmium-based quantum dots, and enhancing charge extraction mechanisms to maximise efficiency [51].

Organic photovoltaics (OPVs): Flexible and Versatile: OPVs are another class of emerging solar cell technologies that use organic (carbon-based) materials to convert sunlight into electricity. OPVs are particularly attractive due to their flexibility and lightweight, making them ideal for niche applications such as wearable electronics, flexible displays, and integration into architectural designs [52]. Their ability to be manufactured using roll-to-roll printing techniques offers potential for cost-effective, high-volume production, that can significantly reduce the per-watt cost of solar energy [53].

In terms of stability and efficiency, OPVs currently trail silicon and perovskites, but continued material science research is quickly enhancing their capabilities. The development of new organic materials with improved environmental stability, charge transport capabilities, and light absorption is underway. Narrowing it down to SSA, OPVs can provide versatile and easily deployable solar solutions for temporary shelters, remote sensing, and even as coatings on various surfaces, contributing to decentralised energy access [54].

Pushing the boundaries of mature solar technology with advanced silicon technologies: While novel PV materials are gaining prominence, significant advances in material engineering and optimisation continue to be realised within the established framework of crystalline silicon solar cells [55]. These advancements aim to push the efficiency limits of silicon, reduce material consumption, and enhance durability. Technologies like Passivated Emitter and Rear Cell (PERC) technology, Heterojunction (HJT) cells, and Tunnel Oxide Passivated Contact (TOPCon) cells have significantly improved the efficiency of commercial silicon modules [56]. These innovations focus on reducing recombination losses and improving light trapping within the silicon wafer. Furthermore, advancements in silicon wafer manufacturing, such as thinner wafers and kerfless production, are reducing material costs and energy consumption in the production process [57].

In SSA, advanced silicon technologies can deliver higher power output from a reduced footprint, benefiting both large-scale solar farms with limited land availability and rooftop installations constrained by space. Their improved efficiency also increases electricity yield per panel, thereby reducing balance-of-system costs. Research into advanced doping techniques, surface passivation, and novel metallisation schemes continues to drive incremental but significant improvements in silicon solar cell performance and reliability [58].

Future material directions for solar energy: The ultimate objective of solar energy material innovation is to develop stable, highly effective solar cells that can absorb a wider range of sunlight. As a result, more research is being done on tandem and multi-junction solar cells, which combine various materials (e.g., G. perovskite on silicon) to absorb various solar spectrum wavelengths, surpassing the single-junction cells' theoretical efficiency limits [43]. In commercial applications, these tandem structures could achieve power conversion efficiencies of over 30%, which would significantly lower the level cost of solar-generated electricity [59].

Material innovations continue to drive the development of next-generation solar energy technologies. From the transformative potential of perovskites and quantum dots to ongoing advancements in silicon and organic PVs, these materials are enhancing the efficiency, affordability, and versatility of solar power. In the context of SSA, harnessing these breakthroughs will be critical for accelerating the energy transition, expanding access to electricity, and supporting a sustainable and resilient energy infrastructure. Sustained investment in research, development, and localised manufacturing of advanced PV materials will be essential to fully realise their impact across the region.

7.6 POLICY, REGULATORY, AND INSTITUTIONAL FRAMEWORKS

Strong institutional, legal, and policy frameworks are essential to fostering an atmosphere that will allow solar energy to be widely used and scaled up in SSA. To draw investment, encourage innovation, and guarantee the sector's sustainable growth, these frameworks offer the stability, incentives, and governance structures that are required.

7.6.1 Policy Initiatives

Many SSA countries have recognised the crucial role of renewable energy sources, especially solar energy, in addressing energy poverty and achieving sustainable development. That acknowledgement has led to the development and implementation of several policy initiatives aimed at supporting the development of solar energy. These initiatives often include the setting of ambitious targets for renewable energy, the establishment of supportive legal and regulatory frameworks, and the introduction of financial incentives.

Renewables targets: An increasing number of SACs have included renewable energy targets in their National Energy Programme and National Determined Contributions under the Paris Agreement. They signal a commitment to a shift away from fossil fuels and give investors and developers clear directions. Countries such as South Africa, Kenya, and Morocco have, for example, set major renewable energy targets, driving significant investments in solar and wind projects [60]. Such targets often form the basis of competitive bidding procedures such as auctioning, which have proved effective in reducing renewable energy costs.

Incentive mechanisms: Various incentive mechanisms have been introduced by governments to entice private sector investment. These may consist of:

- **FiTs**: These give project developers revenue certainly by guaranteeing a set price for electricity produced from renewable sources over an extended timeline. Even though FiTs have been replaced by auctions in some nations, they were essential to the early growth of the renewable energy market.
- The financial viability of solar projects can be greatly increased by lowering or eliminating import duties on solar equipment, providing corporate tax holidays, or allowing accelerated depreciation for renewable energy assets [61].
- Development finance institutions and bilateral donors frequently offer grants or concessional loans to support early-stage development and de-risk projects, especially for off-grid and mini-grid solutions in rural areas.

Frameworks for law and regulation: In addition to offering financial incentives, numerous nations have endeavoured to create laws and regulations specifically pertaining to renewable energy. Creating distinct renewable energy legislation, autonomous oversight organisations, and transparent licensing processes are all part of this. In order to lower regulatory risk and promote project development, the goal is to establish a clear and predictable environment. Additionally, efforts are being made to simplify grid connection procedures and expedite permitting processes to cut project costs and timelines.

Mini-grid and off-grid policies: Several nations have created policies to encourage the development of mini-grids and off-grid solar systems, acknowledging the significance of decentralised solutions for energy access. Subsidies for SHS, rules governing mini-grid operators, and ways to incorporate mini-grids into national electrification initiatives are examples of these policies. The quick expansion of the off-grid solar market in nations like Tanzania and Kenya is proof of the effectiveness of these policies.

The way these policy initiatives are implemented, and their efficacy varies throughout the region, despite these encouraging developments. As the ecosystem of solar technology changes quickly, there are still issues with maintaining policy coherence, bolstering institutional enforcement capabilities, and modifying policies. Nonetheless, policies throughout SSA are continuously improving due to growing political will and awareness of solar energy's potential.

7.6.2 Regulatory Challenges

Although the potential of solar energy is becoming more widely acknowledged, regulatory obstacles still stand in the way of its quick deployment and incorporation into SSA's energy systems. These difficulties are frequently caused by out-of-date laws, ambiguity, and a lack of institutional flexibility to adjust to the rapidly changing ecosystem of renewable energy technologies.

Grid integration is a major regulatory obstacle. Many of the current grid's rules and regulations were created for conventional, centralised power generation, making them unsuitable for integrating variable renewable energy sources like solar. Technical difficulties with grid stability and dependability may result from this, and regulatory frameworks frequently do not provide explicit instructions for integrating distributed solar projects with the national grid. Developers may become uneasy and the smooth integration of solar power into the system may be impeded by the lack of reliable grid codes, such as connection and operating codes [62].

Solar project licensing processes can be difficult, drawn out, and opaque, especially for independent power producers and mini-grid operators. Permits, licences, and approvals are obtained through bureaucratic procedures that frequently result in major delays and higher project costs. Sometimes there are unclear legal frameworks for off-grid energy systems, which leads to misunderstandings about operating guidelines and licensing requirements [63]. In remote areas where off-grid solutions are most needed, this regulatory ambiguity can discourage private sector investment and slow down the pace of electrification.

Tariff structures are another major regulatory obstacle. To draw in investment and guarantee the financial sustainability of projects, fair and cost-reflective solar power tariffs must be established. But because of political factors or a lack of knowledge about the full costs of solar generation, including intermittency and storage costs, tariffs may be set below cost-recovery levels in many SSA nations. Growth and investment in mini-grids may be hampered by tariff concerns, specifically those pertaining to their setting and their relationship to the possible national grid's eventual arrival [64]. For investors and developers of solar projects, revenue certainly requires clear and predictable tariff policies as well as procedures for frequent review and adjustment.

Regulatory agencies' institutional capacity to efficiently monitor and control the quickly changing solar industry is also frequently constrained. Regulators might not have the technical know-how, funding, or autonomy necessary to create and implement suitable rules, keep an eye on compliance, and effectively settle conflicts. Building these institutions' capacity and providing technical support are essential to establishing a stable and alluring climate for solar energy investment. To overcome these regulatory obstacles, a concentrated effort is needed to update legal frameworks, expedite administrative procedures, and improve regulatory institutions' capacities to promote the broad use of solar energy throughout SSA.

7.6.3 Capacity Building and Local Ownership

In SSA, substantial investment in capacity building and encouraging local ownership are crucial for the solar energy sector's long-term sustainability and sustainable growth. This entails ensuring that communities have a stake in the creation and administration of solar projects and providing the local workforce with the knowledge and skills needed along the whole solar value chain, from technical know-how to entrepreneurial aptitude.

Skill development, training, and education: A skilled workforce that can install, run, and maintain solar PV systems, mini-grids, and other solar infrastructure is essential given the quick adoption of solar technologies. Those who are knowledgeable about solar technology, safety procedures, and troubleshooting include engineers, technicians, and electricians. For the development of this expertise, technical colleges, vocational training programmes, and on-the-job training initiatives are essential [65]. Young African professionals are being prepared for positions in the renewable energy industry through several initiatives that emphasise practical, hands-on training [66]. The need for skilled labour at the community level is being directly addressed by initiatives in Tanzania and Uganda, for example, which are creating short-course training programmes to qualify youth as solar artisans [67]. These curricula ought to be customised to meet the unique requirements of the regional setting, integrating real-world knowledge and experience. Giving local communities, these skills not only open up job opportunities but also guarantee the sustainability and dependability of solar installations over the long run, lowering dependency on outside knowledge and encouraging local ownership.

Building capacity in areas like project management, financial literacy, and entrepreneurial skills is necessary in addition to technical skills, especially for local companies that distribute, sell, and maintain solar products. The entire ecosystem that supports solar energy can flourish locally thanks to this all-encompassing approach to skill development.

Promoting local ownership: The sustainability and success of solar energy projects depend heavily on local ownership. Greater acceptance, better maintenance, and a more equitable distribution of benefits result from communities and local businesses having a direct say in the design, execution, and management of solar projects. There are several ways to accomplish this.

- **Community-based models**: Giving local communities, the authority to run SHS programmes or mini-grids can improve sustainability and guarantee that energy solutions are suited to their needs and preferences.
- **Requirements for local content**: Policies that promote or require the use of local labour, supplies, and services in solar projects can boost regional economies, generate employment, and strengthen domestic production of solar componentry [68].
- **Support for local businesses**: By offering financial and technical assistance to small and medium-sized businesses operating in the solar industry locally, from distribution and sales to installation and upkeep, the community's economy is built and a sustainable supply chain is established.

- SSA can guarantee that the solar energy transition is propelled not only by outside investment but also by a knowledgeable, capable, and involved local community by emphasising capacity building and encouraging true local workforce and ownership. This will result in more resilient and significant energy solutions.

7.7 RECOMMENDATIONS FOR ACCELERATING THE ENERGY TRANSITION

A coordinated effort on several fronts is needed to accelerate SSA's energy transition and fully realise the continent's enormous solar energy potential. Building on the opportunities and challenges noted, the following suggestions address important areas for intervention.

7.7.1 Strengthening Policy and Regulatory Frameworks

For solar energy to be scaled in SSA, a stable, open, and investor-friendly policy and regulatory environment must be established. Reforms that give developers and financers long-term clarity and predictability are essential because inconsistent policies and regulatory uncertainties have historically discouraged investments.

Simplifying laws to encourage solar energy investments: SSA governments must make the licensing, project development, and grid connection procedures easier and faster. This entails streamlining the permitting process, cutting down on red tape, and making sure that laws are tailored to the particulars of renewable energy projects. Clear and effective regulatory processes will drastically cut project development costs and schedules, increasing the appeal of solar investments. For example, efficiency and transparency can be significantly increased for renewable energy projects by implementing a single-window clearance system [69].

Setting transparent, long-term goals for renewable energy: Investors are strongly encouraged to consider a nation's commitment to solar development when it has clear, ambitious, and long-term goals supported by policies that encourage them. National energy plans should incorporate these goals and periodically review and update them. Goal setting, procurement procedures (including competitive bidding and auctions), and tariff structures (e.g., A. To boost investor confidence, feedback on tariffs, or PPAs) are crucial [70]. Large-scale solar projects require predictable and bankable PPAs to guarantee developers a steady flow of income. Additionally, policies that encourage local content and value additionally, local assembly, or incentives for domestic solar component manufacturing, can support job creation and economic growth in the solar industry.

Effective legislative and regulatory changes will not only encourage more public and private investment but also create a competitive market, which will eventually result in cheaper solar energy prices and easier access for people throughout the region.

7.7.2 Enhancing Technical and Institutional Capacity

Building institutional and technical capabilities throughout SSA is essential to maintaining the expansion of the solar energy industry and guaranteeing its successful integration into national energy systems. This entails making investments in the development of human capital, encouraging innovation and research, and bolstering the capacities of governmental and regulatory organisations.

Investing in training and skill development: A strong solar industry depends on having a skilled workforce. Comprehensive training programmes covering the whole solar value chain—from production and assembly to installation, use, and upkeep of solar PV systems and mini-grids—are necessary for this. To meet industry demands, technical colleges, universities, and vocational training facilities must create curricula that offer practical skills and hands-on experience [71]. To fill this gap, initiatives such as the Energy Transition Academy are made to provide professional networking services and skill development [72]. To maintain stability and dependability when integrating variable renewable energy sources into current grids, specific training for utility staff and grid operators is also necessary.

Research and Development (R&D): Although a large portion of solar technology is currently imported, it is essential for long-term sustainability and innovation to support local R&D capabilities. Supporting academic institutions and research centres in their applied research on locally relevant solar technologies, such as dust reduction, high-temperature performance, and effective storage options, is part of this. The creation of context-specific solutions and the establishment of a knowledge base in the area can be accelerated through cooperative research projects involving academia, business, and foreign partners [73]. Additionally, by fostering local manufacturing and value addition, R&D investments lessen dependency on imports and promote economic expansion.

Strengthening regulatory and governmental bodies: Attracting investment and guaranteeing fair competition in the solar industry depend heavily on successful governance and strong regulatory oversight. To do this, energy ministries, regulatory bodies, and other pertinent government organisations must bolster their institutional capacity. Building these institutions' capacity should concentrate on fostering proficiency in energy planning, policy-making, regulatory design, contract negotiations, and project management. Training on tariff setting, grid interconnection regulations, and competitive procurement best practices for renewable energy projects are all included in this. Governments will be able to effectively enforce laws, manage the energy transition, and establish a more predictable and transparent investment climate with increased institutional capacity [74]. Additionally, fostering data collection and analysis skills within these organisations will facilitate evidence-based decision-making and improved tracking of advancements made towards renewable energy and energy access goals.

SSA can ensure that the solar energy transition is led by local expertise and backed by strong governance frameworks by making strategic investments in these areas and establishing a solid foundation of institutional and human capacity.

7.7.3 Fostering Regional Cooperation

In order to accelerate the energy transition and optimise the advantages of solar energy in SSA, regional cooperation and integration are essential. Working together can help nations overcome their own constraints, maximise the use of available resources, and build a more resilient and integrated energy system.

Improving international cooperation on solar energy projects: Large-scale solar projects may not be as financially feasible in many SSA nations due to their small energy markets and demand. The construction of bigger, more effective solar farms that can supply several nations is made possible by cross-border collaboration. Reductions in expenses, increased project bankability, and economies of scale can result from this. By facilitating the exchange of technical know-how, best practices, and regulatory harmonisation, collaborative efforts can also make it simpler to execute solar projects internationally [75]. Examples of this type of cooperation include the cooperative creation of solar parks and the planned implementation of renewable energy. Within the next ten years, the G20, for example, hopes to support at least three major cross-border projects that will improve energy trade and resilience in Africa [60].

Establishing regional power pools to promote cooperation and energy trade: Regional power pools are essential tools for improving resource allocation and energy security. The SAPP, WAPP, EAPP, and CAPP are among the power pools that are presently operational or under development in SSA [72]. By allowing nations to trade electricity, these power pools help to balance supply and demand over a larger geographic area. This is especially crucial for solar energy, since it aids in intermittency management. During periods of peak generation, nations with an abundance of solar resources can import power from other sources (e.g., the sun) and export excess solar power to neighbouring nations. G. when solar output is low, hydropower, or other renewable energy sources. This connectivity improves the overall stability and dependability of the regional grid and lessens the need for large-scale home storage solutions [76].

To evacuate power from massive solar projects situated in remote locations to demand centres, regional power pools also make it easier to build regional transmission infrastructure. These pools can lower electricity prices, increase investment in renewable energy, and make the energy market more competitive by promoting energy trade. To fully realise solar energy's potential as a regional resource, measures to fortify current power pools and create new ones, as well as attempts to standardise energy laws and regulations internationally, are essential. In addition to quickening the energy transition, this coordinated strategy will promote increased stability and economic integration throughout SSA.

7.8 CONCLUSION

7.8.1 Summary of Key Findings

With a substantial energy access deficit and an urgent need for a sustainable energy transition, SSA is at a turning point in its energy development. Due to favourable climatic conditions and high solar irradiance, this review has brought attention to the region's enormous, but mainly unrealised, solar energy potential. The lack of adequate transmission and distribution infrastructure, the requirement for reliable energy storage solutions to mitigate intermittency, and the substantial financial barriers resulting from high upfront costs and perceived investment risks are some of the major obstacles that prevent widespread solar adoption, despite its potential. Inconsistent policy and regulatory frameworks, social and cultural barriers like a lack of awareness and resistance to new technologies, and environmental issues like water and land use all contribute to the landscape's complexity.

The advantages of solar energy, however, greatly exceed these drawbacks. Because solar energy reduces greenhouse gas emissions and replaces fossil fuels, it plays a key role in enabling a low-carbon energy transition. To alleviate energy poverty and promote socio-economic development through enhanced health, education, and agricultural productivity, it provides a decentralised and quick way to increase access to electricity, especially in rural areas that are underserved. Additionally, the solar industry boosts economic growth by generating jobs in manufacturing, installation, and maintenance, as well as by supporting creative business models like PAYG solar. Importantly, solar energy increases resilience and energy security by lowering reliance on unstable imported fossil fuels and decentralising power generation, all the while bolstering resilience against the effects of climate change.

7.8.2 Future Outlook

SSA has a bright future for solar energy, but it will require coordinated and coordinated efforts from local communities, governments, businesses, and international organisations. With its wealth of solar resources, the area is well positioned to lead the way in a fair and just global energy transition, promote sustainable economic growth, and provide universal access to energy.

The reliability and cost-effectiveness of solar power will continue to be enhanced by technological advancements like floating solar installations, more effective PSCs, and sophisticated energy storage systems. More energy management optimisation and the integration of various solar solutions into national and regional grids will be made possible by the growing use of smart grids and digitisation. Solar energy will become an even more attractive and affordable solution to the region's expanding energy needs because of these developments.

A persistent dedication to fortifying institutional and technical capacity, promoting regional cooperation, and fortifying policy and regulatory frameworks is necessary to realise this future. Governments must keep creating clear incentives for solar development, streamlined regulatory procedures, and stable, transparent policies that draw investment. By making investments in local R&D, training, and skill development, a strong local ecosystem that can support the solar energy transition will be created. Additionally, bolstering cross-border energy trade and regional power pools will improve grid stability, maximise resource utilisation, and hasten the rollout of massive solar projects throughout the continent and beyond.

In the end, how solar energy develops in SSA will depend on how determined everyone is to get past current obstacles and realise the revolutionary potential of this abundant and clean resource. SSA's sustainable, solar-powered future is not just an ideal; it is a realistic necessity that necessitates prompt, concerted action from all parties involved, opening the door to a more prosperous, healthy, and electrified continent.

REFERENCES

1. Fiona Lambe, Jacqueline Senyagwa, Marie Jürisoo and Hannah Wanjiru (2015). *Clean, Affordable Cooking Energy for African Households.* https://www.sei.org/publications/bringing-clean-safe-affordable-cooking-energy-to-households-across-africa-an-agenda-for-action/
2. Kaihao Cai, Thibault Lemaire, Andrea Medici, Giovanni Melina, Gregor Schwerhoff and Sneha D Thube (2024). *Harnessing Renewables in Sub-Saharan Africa.* https://www.elibrary.imf.org/view/journals/066/2024/005/article-A001-en.xml
3. Low Carbon Power. (n.d.). *Electricity in Sub-Saharan Africa in 2022.* https://lowcarbonpower.org/region/Sub-Saharan_Africa
4. Climate Analytics. (2022). *Renewable Energy Transition in Sub-Saharan Africa.* https://climateanalytics.org/press-releases/renewable-energy-transition-in-sub-saharan-africa
5. Gracelin Baskaran and Sophie Coste (2024). *Achieving Universal Energy Access in Africa amid Global Decarbonization.* https://www.csis.org/analysis/achieving-universal-energy-access-africa-amid-global-decarbonization
6. Maklewa Agoundedemba, Chang Ki Kim and Hyun-Goo Kim (2023). *Energy Status in Africa: Challenges, Progress and ....* MDPI. https://www.mdpi.com/1996-1073/16/23/7708
7. IEA. (n.d.). *Key Findings – Africa Energy Outlook 2022 – Analysis.* https://www.iea.org/reports/africa-energy-outlook-2022/key-findings
8. Tracking SDG7. (n.d.). *Chapter 2- Access to Clean Fuels and Technologies for Cooking.* https://trackingsdg7.esmap.org/data/files/download-documents/sdg7-report2022-ch2-access_to_clean_cooking.pdf
9. IEA. (n.d.). *Access to Clean Cooking – SDG7: Data and Projections.* https://www.iea.org/reports/sdg7-data-and-projections/access-to-clean-cooking
10. Mohamed A. Eltahir Elabbas, Laurens de Vries and Aad Correljé (2023). African power pools and regional electricity market design. Vol 105, 103291. https://www.sciencedirect.com/science/article/pii/S2214629623003511
11. Vedaste Ndayishimiye, Geofrey Bakkabulindi, Emmanuel Wokulira Miyingo, Emmanuel Wokulira Miyingo and Deborah Nibagwire (2025). *Power Generation Overcapacity Selected Sub-Saharan Africa Frontiers in Energy Research.* https://www.frontiersin.org/journals/energy research/articles/10.3389/fenrg.2025.1549844/full
12. Dave Jones (2025). *Solar Power in Africa The First Evidence of a Take-Off in Solar in Africa.* https://ember-energy.org/latest-insights/the-first-evidence-of-a-take-off-in-solar-in-africa/
13. Maklewa Agoundedemba, Chang Ki Kim and Hyun-Goo Kim (2023b). Energy status in Africa: challenges. *Progress and Sustainable Pathways Retrieved from Energies,* Vol. 16(23), 7708. https://doi.org/10.3390/en16237708

14. World Economic Forum. (2022). *Africa is Leading the Way in Solar Power Potential.* https://www.weforum.org/stories/2022/09/africa-solar-power-potential/
15. Vivien Kizilcec and Priti Parikh 2020). Solar home systems: a comprehensive literature review for Sub-Saharan Africa. *Energy for Sustainable Development*, Vol. 58, 78–89. https://doi.org/10.1016/j.esd.2020.07.010
16. Oluleke O. Babayomi, Babatunde Olubayo, Iheanacho H. Denwigwe, Tobiloba E. Somefun, Oluwaseye Samson Adedoja, Comfort T. Somefun, Kevwe Olukayode and Amarachi Attah (2023). A review of renewable off-grid mini-grids in Sub-Saharan Africa, *Frontiers in Energy Research*, Vol. 10. https://www.frontiersin.org/journals/energy-research/articles/10.3389/fenrg.2022.1089025/full
17. Mathilde Brix Pedersen and Ivan Nygaard (2018). *The Case of Private Solar Mini-Grid Development.* https://doi.org/10.1016/j.erss.2018.03.010
18. UNEP CCC. (n.d.). *Rural Electrification through Solar Powered Mini-Grids.* https://unepccc.org/project/rural-electrification-through-solar-powered-mini-grids-a-study-of-private-sector-models-in-kenya/
19. ENGIE Energy Access. (n.d.). *Mini-Grids.* https://engie-energyaccess.com/mini-grids/
20. Power Technology. (2019). *The Biggest Solar Projects in Africa.* https://www.power-technology.com/features/the-biggest-solar-projects-in-africa/
21. CLG Global. (2025). *Powering Africa's Future: The Rise of Utility-Scale Solar ....* https://clgglobal.com/powering-africas-future-the-rise-of-utility-scale-solar-and-wind-projects/
22. Anne Louise Koefoed and Sujee Selvakkumaran (2025). *Costly Capital: Money for Green Megawatts in Sub-Saharan Africa.* https://www.dnv.com/energy-transition/costly-capital-money-for-green-megawatts-in-sub-saharan-africa/
23. World Bank. (2024, November 30). *African Solar Looks beyond Donor Nations for Investment.* https://www.pv-magazine.com/2024/11/30/african-solar-looks-beyond-donor-nations-for-investment/
24. Jennifer Enyonam Anku (2025). *Barriers to Renewable Energy Adoption in Sub-Saharan Africa: a Stakeholder Perspective SSRN.* https://papers.ssrn.com/sol3/Delivery.cfm/5087846.pdf?abstractid=5087846&mirid=1
25. NEPAD. (2024). *Enhancing Access to Electricity through Renewable Energy.* https://www.nepad.org/blog/empowering-africa-enhancing-access-electricity-through-renewable-energy
26. Bamidele Paul Akinbamiwa, Raymond Ere and Shareef Gbadamosi (2025). The impact of solar energy expansion on rural electrification in Africa. *International Journal of Engineering and Modern Technology*, Vol. 11 (3). https://iiardjournals.org/get/IJEMT/VOL.%2011%20NO.%203%202025/The%20Impact%20of%20Solar%20Energy%20186-208.pdf
27. RES4Africa. (2023). *Africa's Energy Future is Renewable.* https://res4africa.org/wp-content/uploads/2023/06/Africas-Energy-Future-is-Renewables-Flagship2023.pdf
28. David Ockwell, Rob Byrne, Ulrich Elmer Hansen, James Haselip and Ivan Nygaard (2024). The uptake and diffusion of solar power in Africa: socio-cultural and political insights on a rapidly emerging socio-technical transition. Vol 44, 122–129. https://doi.org/10.1016/j.erss.2018.04.033
29. Hillary Chanda, Eugene Mohareb, Michael Peters and Chris Harty (2025). Community-led solar energy technology adoption in rural Zambia: the role of observational learning and neighbor influence. *Energy Research & Social Science*, Vol. 122,103972
30. Maryam K. Abdelrazik, Sara E. Abdelaziz and Mariam F. Hassan (2022). Climate action: prospects of solar energy in Africa. *Energy Reports*, 8, 11363–11377.
31. CLG Global. (2025). *Powering Africa's Future: The Rise of Utility-Scale solar and Wind Projects.* https://clgglobal.com/powering-africas-future-the-rise-of-utility-scale-solar-and-wind-projects/

32. World Steel. (n.d.). *Closing the Energy Gap in Sub-Saharan Africa.* https://worldsteel.org/media/steel-stories/infrastructure/closing-energy-gap-sub-saharan-africa/
33. Erik Deitersen (2025). *Top 10 Solar Projects in Africa.* https://www.abiq.io/top-10-solar-projects-in-africa/
34. Gondia Sokhna Seck, Michael Renner and Holle Wlokas (2025). *Large Scale Wind and Solar Projects in Sub-Saharan Africa.* https://doi.org/10.13140/RG.2.2.15787.14887
35. CSIS. (2024). *Achieving Universal Energy Access in Africa amid Global Decarbonization.* https://www.csis.org/analysis/achieving-universal-energy-access-africa-amid-global-decarbonization
36. The Borgen Project. (2024). *Harnessing Solar Energy in Sub-Saharan Africa.* https://borgenproject.org/energy-in-sub-saharan-africa/
37. Yibeltal Wassie and Muyiwa Adaramola (2021). Socio-economic and environmental impacts of rural electrification with Solar Photovoltaic systems: evidence from southern Ethiopia. *Energy for Sustainable Development*, Vol. 60, 52–66. https://doi.org/10.1016/j.esd.2020.12.002
38. Ali Akbar Firoozi and Mohammad Reza Maghami (2025). Harnessing photovoltaic innovation: advancements, challenges, and strategic pathways for sustainable global development. *Energy Conversion and Management: X*, Vol. 27, 101058. https://www.sciencedirect.com/science/article/pii/S2590174525001904
39. Abdussalam Ali Ahmed, Yasser Fathi Nassar, Yasser Abdulgader Alsharif and Muftah Abuali (2025). *Solar Energy: Principles, Technologies, and Applications in Modern Society.* European Academy of Science and Development.
40. Kaihao Cai, Thibault Lemaire, Andrea Medici, Giovanni Melina, Gregor Schwerhoff and Sneha D Thube (2024). *Harnessing Renewables in Sub-Saharan Africa.* https://www.elibrary.imf.org/view/journals/066/2024/005/article-A001-en.xml
41. Business Magazine UAE. (2025). *AI and IoT in Solar Energy: Optimizing Performance and Energy Management.* https://businessmagazineuae.com/ai-and-iot-in-solar-energy-optimizing/
42. Pei Zhang, Huanggen Yang, Qi Zheng, Guochao Nie, Abdullah Yahya and Abdullah Alzahrani (2025). *International Journal of Hydrogen Energy*, Vol. 106, 1088–1113.
43. Kaihao Cai, Thibault Lemaire, Andrea Medici, Giovanni Melina, Gregor Schwerhoff and Sneha D Thube (2024). Harnessing renewables in Sub-Saharan Africa. *IMF eLIBRARY.* https://www.elibrary.imf.org/view/journals/066/2024/005/article-A001-en.xml
44. World Bank Group Report (2025). *Powering Africa: The Transformational Impact of Regional Energy Projects in West Africa.* https://projects.worldbank.org/en/results/2025/02/06/powering-africa-the-transformational-impact-of-regional-energy-projects-in-west-africa
45. David Phaho and Stephen Mathetsa (2025). *Africa's Power Pools: What the G20 Can Do to Help Countries Share Electricity.* The Conversation. https://theconversation.com/africas-power-pools-what-the-g20-can-do-to-help-countries-share-electricity-257561
46. Babatunde Odetayo and Michael Walsh (2021) A policy perspective for an integrated regional power pool within the Africa Continental Free Trade Area, *Energy Policy*, Vol. 156(112436). https://doi.org/10.1016/j.enpol.2021.112436
47. African Climate Foundation Report (2022). *Energy Transitions, Regional Electricity Markets and Industrialisation Prospects in Africa.* https://africanclimatefoundation.org/wp-content/uploads/2022/09/800644-ACF-05_Clean-energy-industrialisation-WEB.pdf
48. S. Saha and M. A. Islam (2025). Potential for nano-enhanced molten salts in solar energy storage. *Renewable and Sustainable Energy Reviews*, Vol. 210(115217). https://doi.org/10.1016/j.rser.2024.115217
49. Prabhat Ranjan Mishra (2025). Simple salt additive gives perovskite solar cells 22.3% efficiency. *Interesting Engineering.* https://interestingengineering.com/energy/salt-boosts-perovskite-solar-cells-efficiency

50. Jun Qu, Wenfeng Zhang, Yangdi Chen, Huiyao Zhao, Yanbei Wei, Rui Zhou, Hongjie Wan, Jie Deng, Dongyong Fan, Yuchen Luo and Haijin Li (2025). Multifunctional amine salt additive for enhancing the performance of perovskite and perovskite/silicon tandem solar cells. *Chemical Engineering Journal*, Vol 515(16396). https://doi.org/10.1016/j.cej.2025.163967
51. Panneerselvam Priya and Albert Alexander Stonier (2025). Emerging innovations in solar photovoltaic (PV) technologies. *Energy Reports*, Vol. 14, 216–242. https://doi.org/10.1016/j.egyr.2025.06.003
52. Asadullah Muhammad Hossain Saad (2025). Next-generation solar cells: advancements in materials, architectures, and system integration for a sustainable energy future. *American Journal of Innovation in Science and Engineering,* Vol. 4(2). https://orcid.org/0009-0008-5256-6299
53. Abdelrahman O. Ali, Abdelrahman T. Elgohr, Mostafa H. El-Mahdy, Hossam M. Zohir, Ahmed Z. Emam, Mostafa G. Mostafa, Muna Al-Razgan, Hossam M. Kasem and Mohamed S. Elhadidy (2025). Advancements in photovoltaic technology: a comprehensive review of recent advances and future prospects. *Energy Conversion and Management: X*, Vol. 26(100952). https://doi.org/10.1016/j.ecmx.2025.100952
54. Abniel Machín and Francisco Márquez (2024). Advancements in photovoltaic cell materials: silicon, organic, and perovskite solar cells. *Materials,* Vol. 17(5), 1165. doi: 10.3390/ma17051165
55. Usman Noor (2004). *Revolutionizing Solar Energy with Quantum Dot Solar Cells.* 8M Solar. https://8msolar.com/revolutionizing-solar-energy-with-quantum-dot-solar-cells/
56. Shailesh Narain Sharma, Pooja Semalti and Bhawna A. Srinivas Rao (2025). Pioneering advancements in quantum dot solar cells: innovations in synthesis and cutting-edge applications. *Current Opinion in Colloid & Interface Science*, Vol. 77, 101905.
57. Matt Beard, Joe Berry and Dan Friedman (2025). *Photovoltaics Chemistry and Nanoscience Research NREL.* https://www.nrel.gov/chemistry-nanoscience/photovoltaics
58. Neeraj Kant and Pushpendra Singh (2022). Review of next generation photovoltaic solar cell technology and comparative materialistic development. *Materials Today Proceeding*, Vol. 56, Part 6, 3460–3470. https://doi.org/10.1016/j.matpr.2021.11.116
59. Mohammed Farji (2021). Development of photovoltaic cells: a materials prospect and next-generation futuristic overview. *Springer Brazilian Journal of Physics*, Vol.51, 1916–1928.
60. Samy Almosni, Amaury Delamarre, Zacharie Jehl, Daniel Suchet, Ludmila Cojocaru, Maxime Giteau, Benoit Behaghel et al. (2018). Material challenges for solar cells in the twenty-first century: directions in emerging technologies. *Science and Technology of Advanced Materials*, Vol. 19(1), 336–369. https://doi.org/10.1080/14686996.2018.1433439
61. Ankit Mathur (2025). Next-generation solar cell market to soar to US$12.95 billion by 2031, driven by rising solar installations and renewable energy investments – the insight partners.
62. Sadiq Jamal, Aimal Daud Khan and Adnan Daud Khan (2020) High performance perovskite solar cell based on efficient materials for electron and hole transport layers. *Optik n Journal*, Vol. 218, 164787. https://doi.org/10.1016/j.ijleo.2020.164787
63. IEA Energy Outlook (2023). *Key Findings – Africa Energy Outlook 2022 – Analysis.* https://www.iea.org/reports/africa-energy-outlook-2022/key-findings
64. Bamidele Paul Akinbamiwa, Raymond Edward Ereh, Shereef Pelumi Gbadamosi and Akinbobola Sunday Oluwagbem (2025). The impact of solar energy expansion on rural electrification in Africa, *International Journal of Engineering and Modern Technology (IJEMT)*, Vol. 11(3). www.iiardjournals.org
65. RES4Africa. (2023). *A Review of Electricity Policy and Regulatory Frameworks in Africa.* https://res4africa.org/wp-content/uploads/2023/06/A-review-of-electricity-policy-and-regulatory-frameoworks-in-Africa.pdf

66. CLG Global. (2025). *Off-Grid Revolution: Expanding Energy Access in Africa.* https://clgglobal.com/off-grid-revolution-expanding-energy-access-in-africa/
67. Green Mini Grids. (n.d.). *Legal and Compliance, Green Mini Grids.* https://greenminigrid.afdb.org/how-it-works/help-desk-developers-and-operators/legal-and-compliance
68. PV Magazine. (2024). *Solar PV Technical and Vocational Training in Africa.* https://www.pv-magazine.com/2024/02/05/solar-pv-technical-and-vocational-training-in-africa-best-practices-from-the-field/
69. ECREEE. (n.d.). *Capacity Building and Certification Program.* https://www.ecreee.org/capacity-building-and-certification-program/
70. Faustine Msangira (2003). *Enabling Youth Empowerment in Solar Energy.* VET Toolbox. https://vettoolbox.eu/actions/enabling-youth-employment-in-solar-energy-in-tanzania/
71. Glory Oguegbu (2024). *Africa's Renewable Energy Revolution: Will the Continent Lead or Lag in Clean Technology Manufacturing?* https://gloryoguegbu.com/africas-renewable-energy-revolution-will-the-continent-lead-or-lag-in-clean-technology-manufacturing/
72. Wikus Kruger and Olakunle Alao (2024) *Driving Growth: Effective Renewable Energy Tendering in Africa GET.transform.* https://www.get-transform.eu/wp-content/uploads/2024/05/Effective-RE-Tendering-in-Africa_May24.pdf
73. Vanessa Nakate, Ujunwa Ojemeni, Leo Roberts, Matt Webb and Sean Rai-Roche (2025). *From Addis Ababa to Belém: Unlocking Africa's Energy Future.* https://www.e3g.org/news/from-addis-ababa-to-belem-unlocking-africa-s-energy-future/
74. RES4AFRICA (2024). Solar PV technical and vocational training in Africa. *PV Magazine.* https://www.pv-magazine.com/2024/02/05/solar-pv-technical-and-vocational-training-in-africa-best-practices-from-the-field/
75. Africa SDG (2025). *Energy Transition Academy: Empowering Energy Professionals to be Leaders in Clean Energy Locally and Climate Action Globally.* RMI. https://rmi.org/our-work/africa-islands-se-asia/energy-transition-academy/
76. Benedict Winchester, Abdullah M. Maghrabi and Christos N. Markides (2024). *RS–DFID Africa Capacity-Building Initiative Programme Grant: Harnessing Unsteady Phase-Change Heat Exchange in High-Performance Concentrated Solar Power Systems.* Royal Society Publishing. https://doi.org/10.1098/rsfs.2023.0059

Section III

Digitalization in Materials for Net-Zero Applications

8 Digital and Material Innovations for Power-to-X Technologies

Joseph X. F. Ribeiro, Aliyu M. Aliyu, Edward Antwi, Flavio Odoi-Yorke, and Benjamin Anim

8.1 OVERVIEW OF POWER-TO-X TECHNOLOGIES

Power-to-X (PtX) technologies generally describe a collection of transformative technologies capable of converting surplus electrical power, usually from renewable energy (RE) sources (e.g., solar or wind) and biomass [1] to various forms of energy or materials. Power generation from RE sources is particularly key because of the global effort to transition from fossil fuels to clean energy [1]. The "X" represents the various products generated, which include gas, liquid, heat, or chemicals [2,3]. In addition to water utilised for electrolysis, PtX requires other substances to produce the desired chemicals and synthetic fuels. For instance, carbon dioxide is required for methane and methanol production. Carbon dioxide (CO_2) is obtained from local renewable sources including carbon capture from the atmosphere (or from carbon dioxide-intensive industries such as cement factories) or biomass sources (biofuel, biogas plants). Ammonia production, on other hand, requires nitrogen [1]. PtX technologies have grown significantly globally. Projections indicate the technology will continue to grow with a potential to play a significant role in future [4].

The aim of this chapter is to provide an overview of the state of PtX technologies ranging from power to hydrogen, gas, liquids, chemicals, heat, and the materials needed to deliver them. Their status in various jurisdictions globally will be explored to give an up-to-date assessment in Europe, North America, Australia, and Asia. Finally, an outlook for PtX will be given, and the consensus is that PtX technologies are moving in a generally positive trajectory with strong growth expected in the years ahead. Meeting country-specific climate change goals (e.g., being net zero) and advancements including artificial intelligence (AI) and Internet of Things (IoT) are expected to drive PtX as well as many other industrial sectors (Figure 8.1).

8.1.1 Power-to-Hydrogen (PtH$_2$)

Hydrogen possesses the advantage of being a flexible energy carrier which facilitates energy production with low to insignificant impact on the environment [5]. Currently, hydrogen is used in the chemical industries [6] and in the agricultural sector where it is employed for fertiliser production [7]. It is on the verge of being utilised in the

DOI: 10.1201/9781003582304-11

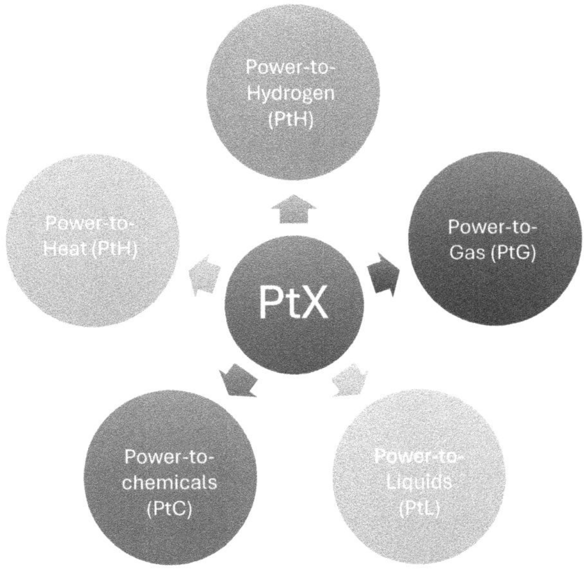

FIGURE 8.1 Main classification of PtX technologies.

transportation sector, which is responsible for a third of global carbon dioxide emissions. Enormous research efforts are being channelled towards production of hydrogen for the transportation sector with the view to cut down emissions by a third [6].

Hydrogen is produced via several methods. Hydrogen produced is usually colour-coded to indicate the method used for production. Green hydrogen, the cleanest, is produced using RE and is characterised by zero emissions. Grey hydrogen can be obtained by every hydrogen production method employing hydrocarbons such as steam reforming process (SMR). When SMRs are integrated with carbon dioxide capture techniques, the hydrogen produced is designated with the colour blue. Pink hydrogen is produced using nuclear energy [1,5,8]. A comprehensive review of the types of hydrogen production is presented by Jain et al. [9].

Hydrogen is produced from fossil fuels, biomass, and via water dissociation [10]. The key technologies employed include partial oxidation of heavy oils, steam reforming [6], coal gasification, and pyrolysis. These technologies require catalysts in specific reactors with temperatures reaching at least 1500°C. Biochemical and thermochemical (liquefaction, gasification, and pyrolysis) processes are the main techniques used in hydrogen production from biomass [6,11,12]. These technologies are still nascent but hold great potential for sustainable hydrogen production [6]. The main technologies employed in water dissociation include electrolysis, photo-electrolysis, and water thermolysis. Water dissociation techniques require energy which can be obtained from fossil fuels, RE, and nuclear energy [6]. Declining costs of RE is enhancing interest in utilisation of RE technologies for hydrogen production [13]. For water electrolysis, the main electrolysers include polymer electrolyte membrane (PEM) (efficiency = 50%–75%), solid oxide electrolyte electrolyser (SOEC) (efficiency = >80%), alkaline water electrolyser (efficiency = 50%–75%), and anion exchange membrane (AEM) (efficiency = 55%–60%) which is an emergent technology. The PEM technology is commercialised and produces hydrogen with purity of 99.99%. The AEM technology is a mature technology with hydrogen purity of between 99.5% and 99.9%. The SOEC can produce hydrogen with 99.9% purity; however, production remains at lab scale. The AEM generates a hydrogen purity of 99.9% but remains at research and development stage [5]. Global installed capacity of water electrolysis for hydrogen production reached approximately 300 MW by end of 2020 with the alkaline electrolyser being the most widely utilised (61% of the total installed capacity, while PEM accounted for 31%). Projections indicate that installed electrolyser capacity could reach 54 GW by 2030 [8]. A comprehensive review of electrolyser technologies is provided by Gomez and Castro [8].

Hydrogen finds application in several sectors, mainly the chemical industry. It is used in fuel cells [14], for refining, in industrial applications (ammonia, methanol, and steel production) [6], residential applications (cogeneration, which involves producing both heat and electricity simultaneously) [8], energy production, and more recently employed in the transport sector, particularly in electric vehicles [10]. Latvia is the only country mainly using hydrogen for transport. Top five countries with the highest hydrogen demand include Germany (1.74×10^6 Mton year^{-2}), the Netherlands (1.30×10^6 Mton year^{-2}), Poland (7.74×10^5 Mton year^{-2}), Spain (5.55×10^5 Mton year^{-2}), and Italy (5.19×10^5 Mton year^{-2}) [5].

After production, hydrogen is stored in several ways: as compressed gas, in underground storage, as liquid hydrogen and solid hydrogen (metal hydrides, complex hydrides, chemical hydrides, and adsorbents) [8,14]. While hydrogen remains promising, key challenges to its scaling up include low production efficiency, high cost of production [8], inadequate safety storage, infrastructure development, and policy and regulation [14].

8.1.2 Power-to-Gas (PtG)

PtG refers to the chemical storage of electrical energy in the form of gases (methane or hydrogen). It can also be defined as utilisation of surplus electrical energy from renewable power sources for hydrogen production in an electrolyser and the optional synthesis of methane or other hydrocarbons from hydrogen and carbon dioxide [15]. PtG presents several advantages including demonstrating the potential of long-term storage of electrical energy and presenting a path for a more resource-efficient and flexible energy system. Again, it enables an accelerated integration and continued realisation of low- or zero-emission technologies (e.g., solar PV and wind), thereby facilitating achievement of both climate and energy policy objectives. Further, the technology introduces innovations by providing new options in energy transport by shifting the load from the power grids to gas grids [15,16]. While most of the process steps within a PtG system are not new and are mature, electrolysis needs to be aligned to the requirements of the PtG technology because RE or excess electricity are usually temporarily available. Relevant electrolysis technologies employed in PtG have been discussed in 1.1. In this technology, the hydrogen produced is combined with carbon dioxide to produce methane. The process, methanation can be implemented via two concepts, catalytic and biological [17]. The methane produced can be used in the industry, transport, electricity, and heat sectors respectively. In 2018, at least 106 PtG projects were documented in Europe, most of them located in Germany [17].

8.1.3 Power-to-Liquids (PtL)

PtL technologies comprise utilisation of renewable electricity to produce synthetic liquid fuels (e.g., kerosene for aviation) [18]. The process involves hydrogen production through electrolysis using RE after which the hydrogen produced is combined with carbon dioxide, using the Fischer–Tropsch process or methanol synthesis, to produce the liquid hydrocarbons [19,20]. The fuels can be utilised in several applications including aviation, transportation, and heating. Production of liquid fuels is essential for several reasons. First, the liquids produced are climate-neutral and contribute to global climate goals. Further, compared to biofuels, PtL achieves higher yields and energy obtained from RE sources. Again, water requirements for PtL are also significantly lower. PtL is also perceived as a key technology to enable a fully regenerative, sustainable, post-fossil fuel supply of aviation in the longer run, while avoiding the potential risks and adverse side effects through the energetic use of cultivated biomass and in land-use. PtL can be produced from concentrated renewable CO_2 sources using established industrial processes with relatively high technology readiness. Individual processes may be applied on large scale; however, PtL full-system integration is recently progressing due to demonstration plants available in Iceland, Finland, Germany, Spain, Switzerland, and Norway. Between 2021 and 2022, it was anticipated that a demonstration plant for PtL jet fuel would be built in Stade, Germany [21]. As of 2015, improved processes for carbon capture for the air and high-temperature electrolysis were at industrial scale demonstration level in Canada and Norway. It was anticipated that the effort could enhance the production potential and efficiency [21].

8.1.4 Power-to-Chemicals (PtC)

PtC refers to utilisation of electricity, particularly from RE sources (solar, wind, or hydropower), to produce valuable chemicals or fuels. This comprises one of the approaches to decarbonise the chemical industry and facilitating reduction of reliance on fossil fuels [22]. Hydrogen is produced via electrolysis [3] and further utilised as feedstock to synthesise desired chemicals including ammonia, methanol, e-fuels (gasoline, diesel), and building blocks for plastic production, among others [23]. Despite the benefits, the technology experiences challenges. First, the technologies (electrolysis and other PtC-related technologies) are expensive [24]. Further, they may require additional development for scaling up [25]. Also, high efficiency must be ensured in converting electricity to chemicals, and vice versa. Additionally, existing infrastructure may have to be modified, or new ones developed for hydrogen or synthetic fuels production.

8.1.5 Power-to-Heat (PtH)

PtH comprises relevant technologies that convert electrical energy into heat (e.g., electrode boilers [26], heat pumps), often utilising excess or renewable electricity. These technologies play a crucial role in stabilising energy grids by absorbing surplus power and offer significant boost to energy transition [27]. Advantages obtained from these technologies include efficient production of (largely CO_2-free) heat, increased flexibility particularly for cases that combine PtH with thermal energy storage, increased utilisation of RE resources, and reduced electricity costs [26].

8.2 THE ROLE OF PTX IN THE RENEWABLE ENERGY ECOSYSTEM

PtX technologies are crucial to the RE ecosystem for several reasons. They enable integration of RE technologies. Two challenges encountered during generation of power from RE sources (e.g., solar and wind) include storage of surplus energy and intermittency. PtX technologies resolve these challenges by utilising excess RE for production and storage of hydrogen and other carriers. To balance the intermittency of RE sources (when combined in power plants) and ensure a more reliable energy supply, hydrogen is converted to power [4]. It also introduces flexibility to power production, in that power production can be implemented in difficult-to-reach areas where micro-grids can be introduced and also across sectors [4]. Noteworthy is the fact that PtX facilitates significant improvements and increments in clean energy production, hence facilitating RE penetration. Projections indicate that RE sources could contribute significantly to the global transition to net-zero emissions and facilitate mitigation of climate change. PtX technologies also facilitate decarbonisation across sectors. A colossal effort has been dedicated and continues to be dedicated to decarbonise hard-to-decarbonise sectors (e.g., shipping, aviation, and agriculture) of the global economy with the view to limit global warning and provide a solid alternative for fossil fuels [2,4,28]. Additionally, PtX facilitates interconnection of various energy sectors (electricity, heating, transportation, and industry), best known as sector coupling, by creating a pathway for energy exchange and utilisation across these sectors [29]. Additionally, PtX facilitates production of sustainable materials and chemicals. PtX technologies facilitate production of sustainable chemicals (ammonia, methanol) and materials from green electricity (from RE sources) instead of employing fossil-based feedstock [30]. Further, PtX technologies contribute significantly to both circular carbon economies and emissions reduction. Some of the processes integrate carbon capture and storage or carbon capture and utilisation. Countries with PtX technologies gain export opportunities in cases where they produce more synthetic fuels and chemicals than they need. While this contributes to global decarbonisation efforts, global trade could be enhanced significantly [4,31,32] (Figure 8.2).

8.3 DIGITAL MATERIAL INNOVATIONS FOR POWER-TO-X TECHNOLOGIES

Digital material innovations for PtX technologies utilise advanced AI and machine learning (ML) to rapidly accelerate design and discovery of new materials relevant to PtX technologies, particularly catalysts and membranes which are critical for efficient electrolysis processes required for power conversion. AI can be used to predict the catalytic activity of various materials to aid water splitting enabling the design of highly efficient electrocatalysts for production of hydrogen. ML techniques are employed to predict the permeability and possibilities for selection of different membrane materials which lead to development of optimal membranes for efficient gas

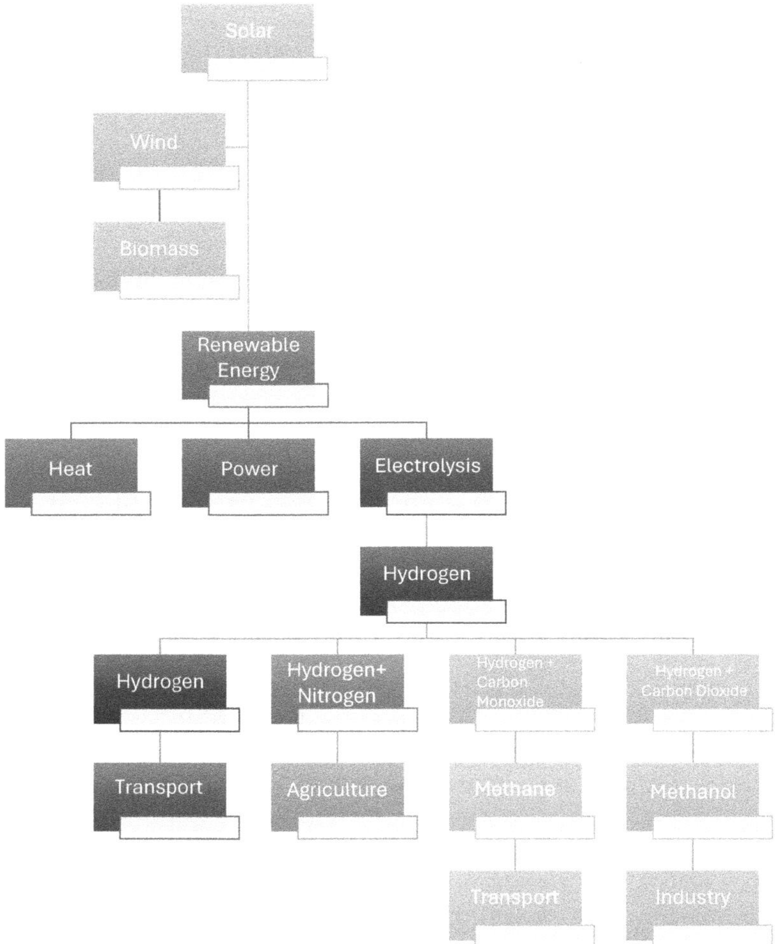

FIGURE 8.2 The PtX value chain.

separation in PtX processes. An example of a material developed using ML is ionic liquids perceived to be cleaner compared to conventional volatile organic solvents and which exhibit promising properties including low vapour pressure, low flammability, and recyclability [33]. These technologies utilise computational models to accurately predict material properties and simultaneously optimise their performance, ultimately leading to enhanced efficiency and durability in PtX applications with the added benefits of reduced time and cost [34–37]. Additionally, Batool et al. [33] report, with many examples, that AI and MLs can be employed in design control, process monitoring, quality control, performance assessment, characterisation data analysis, degradation studies, materials optimisation, and experiment selection. Focussing on fuel cells and electrolysers, the authors posit that ML algorithms can be used to predict and optimise their performance based on compositional and structural parameters. They highlight the ability of the combining MLs with genetic algorithm to further optimise the design and operating parameters to achieve multiple optimisation goals with high accuracy and efficiency.

The advent of smart sensors and the IoT usually embedded in PtX systems enable real-time data (temperature, pressure, flow rates, chemical composition, etc.) acquisition. This achievement enables development of digital twins, virtual replicas of physical processes which utilise live data to optimise reactor performance, reduce material waste, improve efficiency, and support predictive maintenance through simulation and continuous analysis [38,39].

Additive manufacturing (AM), one of the major pillars of Industry 4.0 and application in catalysis [40], enables rapid prototyping of materials for PtX technologies. AM allows fabrication of complex internal devices (e.g., cooling fins) within reactor components that would be challenging or virtually impossible to fabricate via conventional methods. Advantages obtained from utilising the devices obtained include optimised fluid flow, heat transfer, and enhanced reactor efficiency [41]. Nanomaterials (e.g., carbon nanotubes and nanofibers) are crucial for creation of high-performing fuel cell electrodes. This is attributable to their small sizes which create extremely high surface areas for catalysts and efficient gas diffusion, improving the electrochemical reactions that generate electricity. Nanomaterials also enable optimised electrode structures with enhanced porosity, facilitating the continuous flow of gases like oxygen and hydrogen and leading to increased fuel cell efficiency and current generation [42–44]. Integrating nanomaterials into 3D printing materials presents a powerful approach to fabricating complex structures with tailored properties (e.g., improved thermal conductivity, corrosion resistance), particularly in applications like reactor components and energy conversion devices where high performance and specific functionalities are critical [45]. AM with nanomaterials aid creation of porous electrodes with optimised surface area for efficient gas exchange and electrochemical reactions in fuel cells [46]. Further, nanostructured materials incorporated into 3D-printed solar cell structures possess the potential to enhance light absorption and charge carrier transport, improving overall efficiency [47].

Integration of digital platforms, comprising cloud-based data sharing and collaborative databases to accelerate the innovation of PtX materials cannot be overemphasised. The integration fosters rapid real-time and accurate data-sharing, facilitating co-creation, streamlining of value chains, and promotion of the development of technical standards and best practices in industry. The technologies are not limited to industry but cut across academia and even governments who must gather information to provide solutions to pressing challenges in PtX systems [48,49]. The materials cloud comprises a typical example of a digital platform that supports computational material science by integrating various software frameworks and data resources to foster open science and collaboration [50].

8.3.1 Novel Materials Developed for PtX Systems

Advances in materials utilised in PtX systems are increasingly targeted at addressing fundamental challenges of conventional technologies including reliance on rare and expensive materials, high energy consumption, and lack of efficiency and stability, among others. Contemporary research focusses on enhancing the performance of catalysts, electrolytes, and membranes to make production of green hydrogen and other e-fuels scalable and affordable [51–53]. New catalysts being explored include copper-based catalysts (modified copper and its alloys), single-atom catalysts (SACs), metal-free carbon catalysts (carbon materials doped with non-metallic heteroatoms like nitrogen, sulphur, or boron), and bimetallic and alloy catalysts. Copper-based catalysts have become important because of their ability to produce high-value products like ethylene and ethanol from carbon dioxide [54–56]. SACs utilise isolated, single metal atoms dispersed on a support material like nitrogen-doped carbon, which maximises use of metal resources and significantly reduces cost [57,58]. Metal-free catalysts have demonstrated high selectivity for production of CO. Desirable properties they possess like surface area and chemical stability make them a low-cost alternative to metal catalysts [59,60]. Bimetallic and alloy catalysts (e.g., Cu-Sn, Cu-Ag, and Cu-Zn alloys) have demonstrated enhanced activity and stability for reduction of carbon dioxide [61–63].

Advances in membrane development and manufacturing are increasingly focussing on producing more durable, efficient, and cost-effective alternatives to standard perflourosulfonic acid (PFSA) membranes like Nafion. Recent innovations include development of AEMs for electrolyser applications offering cheaper and abundant nickel- or iron-based catalysts [64,65]. New materials for solid electrolytes including doped ceria (e.g., gadolinium-doped ceria, GDC) and lanthanum strontium gallium magnesium oxide have demonstrated improved ionic activity and superior performance compared to yttria-stabilised zirconia (YSZ) during high-temperature electrolysis [66,67]. In parallel, success has been achieved utilising biomass fillers within a polyvinyl matrix to produce PEMs. The novel bio-based material offers a cost-effective and more sustainable alternative to traditional PFSA membranes [68].

Solid oxide electrolysis cells (SOECs) deliver high efficiency; however, they suffer from material degradation at high temperatures. Preferable anode materials include perovskite-type oxides (e.g., lanthanum strontium manganite due to their high electrical conductivity and thermal compatibility with YSZ electrolytes). Recently, research is proving that mixed oxide compositions (including scandium-doped variants and perovskite–chromate combinations) could improve catalytic activity and stability [69]. Further, while cermet materials like nickel and yttria-stabilised zirconia (Ni-YSZ) are common, alternative compositions are being studied to improve performance and stability for carbon dioxide and steam co-electrolysis. For instance, Ni-GDC electrodes offer better interfacial charge transfer and improved resistance to coking and sulphur poisoning in some tests, although microstructural changes like Ni agglomeration pose challenges to long-term stability [70]. Material selection also varies with operating temperature: titanium and carbon cloth are widely utilised in low-temperature electrolyser components, including porous transport or gas diffusion layers, for their corrosion resistance. Conversely, high-temperature SOECs depend on chromium-, nickel-, and iron-based alloys to withstand harsh oxidative and thermal environments [51].

8.4 TECHNICAL AND ECONOMIC CHALLENGES

While PtX technologies are set on the path of growth they are saddled with several challenges [71]. PtX technologies suffer from low efficiency and high cost attributable to expensive catalysts requirements (e.g. platinum), energy-intensive processes, and the difficulty of optimising complex chemical reactions [72]. Technologies utilised for electrolysis have relatively low efficiencies (PEM: 50%–75%, SOE:>80%, AWE: 50%–75%, AEM: 55%–60%). A proposed photovoltaic electrolysis system proposes an overall solar-to-hydrogen conversion efficiency of >30% [73]. The capacity of the electrolysers for PtH comprises one of the key challenges in choosing an electrolyser type. The smallest electrolyser types (>5kW) are too small to be utilised for demonstration projects and are perceived suitable for lab activities. The others, under 100kW are also perceived to be inflexible despite their advantages. There are electrolysers with a capacity of 1 MW. Due to the challenges with the capacities, several designers are considering installation of multi-MW electrolysers. Further, a sizeable number of projects are undecided about the electrolyser kind or capacity to utilise. Also, the CAPEX is quite high for electrolysers. For instance, the CAPEX (CHF/kW) for selected electrolysers for 2020 are ALE (650–1500), PEME (800-2300), and SOE (1600-5750). This is projected to reduce in 2030 (ALE, 460-920; PEME, 460-2000; and SOE, 600-4600) [74]. PtM has balance and storage challenges [74]. As previously iterated, the electrolysers also usually require expensive catalysts like platinum. Other technologies with the capacity to generate fuels from hydrogen and carbon dioxide also require further research and development to improve their efficiency [2]. The inherent challenges and cost, coupled with the complexity of technology, limit scale-up to meet planned decarbonisation targets. Resolution of these challenges could enhance up-scaling the technologies.

PtX technologies also require more infrastructure (e.g., pipelines and storage systems) to enable optimal production, utilisation, and distribution. Development of these infrastructure, retrofitting existing infrastructure for the PtX-related purposes, and developing new supply chains can be complex and costly [75]. Despite protocols to ensure high levels of safety in the PtX industry, risk of accidents remains high particularly for hydrogen storage and transportation. Resolved risks could facilitate increased public acceptance of the technologies [76]. Consistent and clear regulations for PtX technologies appear to be lacking negatively affecting their deployment globally. There are also gaps to fill to ensure regulations and policy frameworks can be significantly synchronised globally. Large-scale PtX facilities would require significant land areas. This could potentially lead to conflicts among inhabitants and competition with other areas such as agriculture [76]. There is also a challenge with sustainable feedstock sourcing. PtX heavily relies on water. Reliable and sustainable water supply required for electrolysis could be challenging for areas with water scarcity. Readily available alternatives could include seawater or wastewater. However, desalination and treatment could significantly increase production cost. In addition, utilisation of these alternatives could negatively impact the environment. Carbon dioxide is also essential to some of the PtX pathways. Ensuring a sustainable and affordable carbon dioxide supply is critical [76].

The major economic challenges mitigating the development and growth of PtX technologies comprise high initial costs particularly for electrolysis, synthesis processes, high investment for production, storage, and distribution infrastructure [71]. It is noteworthy to mention that operational costs comprising consumption and maintenance can be significantly high especially in cases where there is insufficient or lack of government subsidies or incentives [71]. PtX technologies are still burgeoning.

As a result, the markets for PtX products are still developing. Supportive policies and regulations are needed to promote market acceptance and accelerate the development of PtX systems [75,77]. Currently, large-scale PtX projects are underway and further cost reductions are expected through economies of scale; however, this requires continued investment and development [75].

8.5 GLOBAL PROJECTS

PtX technologies are being implemented in many countries. Several countries are actively involved in PtX projects with the view to produce green hydrogen and its derivatives using RE energy sources. The projects range from pilot plants to national strategies and research initiatives, and target decarbonisation of various sectors including transportation and chemicals. As of 2020, 220 projects had been identified in 20 European countries with France and Germany leading the charge. At the time, both countries planned to install 500 MW of capacity by 2025. Other very active countries are Denmark and the Netherlands. Bordering the North Sea, the large potential for offshore and onshore wind could guarantee efficient production of hydrogen and other PtX products [78].

Germany appears to be in the lead in green hydrogen production and PtX technology. As part of the country's National Hydrogen Strategy, Germany is actively developing relevant infrastructure (pipelines and refuelling stations) and policies to enhance development. Germany has set aside USD 14.19 billion for infrastructure development, international collaboration, and hydrogen research. For instance, in 2024, the ThyssenKrupp's Nucera project received USD 746 million from the German government to produce electrolysers [79]. PtX Development and KGAL have signed a financing agreed for the construction of one of Germany's largest hydrogen production plants at Hydrogen-Hub in Lubmin, Mecklenburg-Western. The plant with a capacity of 1050 MW and an annual production of approximately 100,000 tonnes will have the first construction phase output of 210 MW [80]. Uniper is transforming Wilhelmshaven into a central energy hub for the import and production of green hydrogen-based energy sources. The European Commission has selected and published the project along 164 other projects in the official journal of the European Union in April 2024 [81]. Germany also hosts the HydrOxy Hub Walsum project which aims to develop, build, and operate an innovative large-scale PEM electrolyser with a capacity of 1572 MW [82]. The PtX Platform and Fund is also a German initiative that targets closing the bankability gap for PtX projects, particularly for developing countries and emerging economies [83].

As of 2021, Japan was leading on hydrogen, targeting applications in transport, heating, and energy production [84]. Japan pursuit of green hydrogen technology is primarily driven by the need to achieve energy security. The country is home to the Fukushima Hydrogen Energy Research (FH2R) Field (10 MW capacity electrolyser) [84], one of the biggest hydrogen production plants in the world. Japan is also investing in hydrogen fuel cell technology, which is being employed in the transportation sector. Toyota has already demonstrated the potential of the fuel cell application in its model Mirai, a wholly hydrogen-powered vehicle [79]. The government of Japan updated its Basic Hydrogen Strategy (June 2023) with the view to encourage green hydrogen business endeavours. The plan aims to invest approximately USD 98.8 billion (JPY 15 trillion) from 2024 to 2029 in nine selected technologies (including fuel cells and water electrolysis devices) [79]. Work on this project has already begun. Hydrogen from this project will be used to generate electricity to be supplied to local consumers in Niigata Prefecture [85]. Japanese oil and gas firms Eneos and Idemitsu Kosan have signed an MoU with utility Hokkaido Electric Power Company (Hokuden) to develop a green hydrogen project with approximately 100 MW of electrolysis capacity by 2030 [86].

Australia is actively pursuing a range of PtX projects, with a strong emphasis of green hydrogen production and export. The thrust is supported by national strategy and investments in RE infrastructure [87]. Utilising its abundant solar and wind resources for hydrogen production, Australia is rapidly becoming a key stakeholder in the green hydrogen industry. The government of Australia revealed plans in to invest USD 4 billion via its Hydrogen Headstart programme. It is anticipated that the programme will provide revenue support for large-scale renewable hydrogen projects accessed via competitive hydrogen production contracts [79,87]. The Australian government has invested more than USD 500 million and set up hubs like the Western Australian Asian Renewable Energy Hub with the intention to export its hydrogen to Asian markets [87]. With more than 20+ GW-scale projects, it appears Australia is leading the world within the development of large-scale green hydrogen/PtX projects. The government is also modifying its LNG infrastructure to export hydrogen [79,88].

The United States of America (USA), through the Department of Energy, has announced that it would provide USD 750 million for 52 projects spread over 24 states. Support for the project will significantly reduce the price of clean hydrogen while maintaining America's lead in the clean hydrogen market [79,89]. The USA also hosts several PtX projects including the European Energy's E-Methanol Plant in Texas which will produce 100,000 tons of e-methanol annually, providing approximately 200 construction jobs [90]. Others include a 200 MW PtX project by StormFisher with a planned capacity of 1.88 GW by 2030 [91], 1 GW wind-solar-PtX project by Nova located in Texas [92], and GreenGo Energy's expansion via projects including battery energy storage systems and Megaton PtX facilities, among others [93].

The thrust for China's rapidly expanding green hydrogen industry is reduction of carbon emissions and mitigation of air pollution. The 14th Five-Year plan includes significant investments into hydrogen infrastructure. In 2023, China invested USD 42.04 billion (300 billion Yuan) in green hydrogen businesses [79]. Projects ongoing in China include the Baotou Green Hydrogen project, Topsoe and Mintal Hydrogen's Green Ammonia project [94], Offshore Solar Hydrogen Farm (Rudong) [95], Green Hydrogen Project (Ningxia) [96], and the Marine Hydrogen-Ammonia-Methanol Project [97].

South Korea is actively implementing several PtX projects focused on green hydrogen and ammonia targeted at power generation and energy storage [98]. It has a target to dominate the green hydrogen sector globally. To achieve this, it has set high goals for infrastructure growth, hydrogen vehicle adoption, and hydrogen generation. The country's implementation of RE, especially in offshore wind power, is helping to boost green hydrogen generation. Utilising projects like the Ulsan Hydrogen Town project, the country is incorporating green hydrogen into daily lives via residential fuel cells and hydrogen-powered buses [79]. Other projects include implementing hydrogen and ammonia co-firing in power plants [98], offshore wind and hydrogen integration (wind power projects by Taean Wind Power, competitive bidding for offshore wind projects – 1.25 GW capacity offered in first half of 2025) [99]. For hydrogen production and infrastructure, South Korea is planning a 107.9 MW hydrogen fuel cell power plant in Gyeongju, scheduled to completion by 2028 [100]. Additionally, KHNP (a subsidiary of Kepco) implementing a nuclear hydrogen project, intends to produce pink hydrogen using low-temperature water electrolysis. The plan aims to produce more than four tonnes of hydrogen per day by 2027 [101]. Again, South Korea is funding development of the world's largest liquid hydrogen carrier [102].

France is utilising its large nuclear power infrastructure as a low-carbon electrical source for hydrogen production. A key objective of the PtX strategy of France is to decarbonise its industrial sector by developing a strong electrolysis sector [103]. Additionally, the country is investing in RE initiatives including solar and wind power. By 2030, France intends to increase hydrogen production, promote heavy-duty mobility and development of a fully integrated competitive sector. France is also investing heavily in the sector. For instance, the country allocated USD 7.53 billion to support all green hydrogen businesses [79]. Other projects include the Sunrhyse project in South of France utilising RE for continuous hydrogen production [104], and Lhyfe's fifth green hydrogen production site under construction in the North of France [105].

Spain is employing its RE sources to become a leader in the green hydrogen market [106]. The government's Hydrogen Roadmap, launched in 2022, presents high infrastructure development and hydrogen production goals. Spain's potential is demonstrated by initiatives like the Iberdrola green hydrogen plant in Puertollano, which produces hydrogen using solar energy. The development of Spain's hydrogen sector is fuelled by financing from the European Union and government support for hydrogen technologies. Spain's geographic location and capacity for RE make it well positioned to be a significant player in the green hydrogen sector [79].

The Netherlands is investing significantly in green hydrogen as part of its energy transition strategy. For instance, since April 2024, the Netherlands has allocated USD 860 million for electrolyser innovations to achieve the 4 GW capacity planned by 2030. An important part of this innovation is taking place in Europe's largest open test facility for hydrogen, the Faraday Innovation Lab in Petten, north of Amsterdam. Additionally, its extensive offshore wind capacity supports the Netherlands' green hydrogen production. Projects like the NortH2 project, which aims to produce green hydrogen using offshore wind power, highlight the country's commitment to green hydrogen [79]. Recently, Spain has approved €1.22 billion ($1.34 billion) in funding for 2.3 GW worth of green hydrogen hub projects in Aragon, Andalusia, Castilla y León, and Galicia [107].

The UK is making significant investments in green hydrogen as part of its 2050 goal to have net-zero carbon emissions. Launched in 2021, the UK government's Hydrogen Strategy sets high infrastructure development and hydrogen generation goals. The UK's large offshore wind capacity helps fuel its green hydrogen generation. The nation's potential is demonstrated by initiatives like the Tees

Valley Hydrogen Hub, which intends to create hydrogen using offshore wind power. Additionally, the development of the hydrogen economy in the UK is fuelled by the government's backing of hydrogen technology and the active involvement of the private sector. The UK's dedication to sustainable energy and hydrogen technology guarantees its position as a frontrunner in the green hydrogen industry [79]. The UK has at least 10 green hydrogen projects running. Other PtX initiatives include the European Energy's Kasso e-methanol facility and exploration of integration of PtX technology into the energy system by Eurowind Energy to support and supplement existing energy sectors [4,28].

As part of the International PtX Hub, PtX Pathways supports the development of sustainable PtX and hydrogen markets as a building block for the energy transition in Morocco, South Africa, and Argentina. The project focuses on the specific needs of the countries and their individual motivations. With funding of €24.7 million, the project supports the responsible ministries for energy or economy in the development of allocation scenarios for PtX, including the analysis of value chains. Jointly with the partners, the team identifies opportunities for business development and derives recommendations for improving the regulatory framework for PtX. In Morocco, a PtL pilot plant demonstrates the entire PtX value chain, paving the way for scaling and capacity development. Good practices, tools, and guidelines are disseminated to other countries via the International PtX Hub and thus fed into the PtX debate in other markets [108].

Chile, El Salvador, and Kenya are exploring PtX projects, with a focus on specific resources like geothermal energy [109]. In Saudi Arabia, the NEOM Green Hydrogen Project, aiming to be the world's largest green hydrogen plant, is under construction. The project has reached 80% construction completion across all sites. The mega-plant is expected to produce up to 600 tonnes per day of carbon-free hydrogen in the form of green ammonia as a cost-effective solution for transportation and industrial sectors globally [110].

Several European countries, including Finland and Norway, are also involved in PtX projects, through EU initiatives. Europe's Innovation Fund's pilot hydrogen auction (IF23 Auction) is funding six projects in various countries: Spain (Hysencia and Catalina), Portugal (Grey2green II and MP2X), Finland (eNRG Lahti), and Norway (SkiGA) [111]. Denmark has developed a new national PtX strategy (2021). New policies have been approved to unlock green fuel production potential across the country. Ammonia is forecasted to be the cheapest long-term option among the hydrogen derivatives [112].

In Africa, Morocco is demonstrating the entire PtL value chain and developing a research and training plant for sustainable aviation and maritime fuels. Additionally, it is developing national hydrogen strategies and exploring utilisation of RE sources for PtX applications [108]. Under the same project, the PtX Hub has showcased the project and hosted knowledge sessions. The country is actively developing national hydrogen strategies and assessing market opportunities for green shipping [108]. Kenya is targeting to employ its abundant resources (especially hydro and geothermal) for PtX projects (sustainable electricity-based materials and fuels for the international market) [113]. A working group has been established to explore installation of a 50 MW electrolyser capacity.

8.6 STRATEGIC OPPORTUNITIES AND OUTLOOK

PtX technologies appears to be at the forefront of energy transition with a strong capacity to transform green electricity to crucial energy carriers including hydrogen, synthetic fuels, chemicals, and possibly other substances yet to be discovered [114]. The technologies present benefits that can be exploited to significantly reduce greenhouse gas emissions while simultaneously supporting green transition [4]. PtX technologies are on the cusp of explosive growth attributable to global shift towards green energy production and sustainable utilisation systems. PtX also present several strategic opportunities. First, they hold key to decarbonise hard-to-abate sectors (e.g., cement, transport). They are essential to facilitate emissions reduction in sectors including cement production, transportation, and agriculture which are challenging to electrify directly [115,116]. Further, PtX enhances the ability to store excess energy (particularly RE in the form of hydrogen or synthetic fuels). The stored energy provides a buffer for fluctuation in RE sources, ensures a stable energy supply, and protects electricity grids by ensuring grid balancing [4]. In addition, PtX technologies also present the opportunity to be scaled (from small to large scale) at various levels of applications. This largely introduces resource efficiency supporting sustainability goals [117]. Additionally, these technologies present the opportunities to create new industries having the potential to rapidly expand, creating new jobs and economic opportunities in various sectors globally which were previously non-existent [117,118]. Araya et al. [1] reported 40 PtX projects generating 1200 full-time jobs in Denmark at the time of publication. PtX is also expanding markets for products generated. For instance, methanol market was reported to be increasing at an average annual growth rate of approximately 7% to an estimated USD 91.53 billion by 2026. The ammonia market is projected to grow from USD 71.98 billion in 2021 to USD 110.93 billion in 2028. The hydrogen market is estimated to grow with a CAGR of 5% until 2030 [1]. PtX provides a crucial thrust to circular economy, facilitating production of fuels and chemicals from waste and by-products. This presents a concrete alternative utilisation of fossil fuels while reducing reliance on virgin resources [119]. Furthermore, significant cost reductions are expected for all PtX technologies until 2050 attributable to scaling and learning effects with the magnitude of the cost reduction differing with each technology and degree of technological maturity [120].

The outlook for PtX technologies is generally positive. Strong growth of the technologies is expected in the years ahead. PtX technologies are perceived as crucial for decarbonising various sectors and achieving net-zero emissions. Technological advancements including AI, IoT, and advanced materials are driving digitalisation in PtX technologies. This integration will lead to increased efficiency, predictive capabilities, and optimised resource management in PtX processes. For instance, AI can accurately analyse data from installed IoT sensors in PtX facilities to optimise energy consumption and predict potential equipment failures. This will translate to improved operational efficiency, reduced downtime, and lower costs [121,122].

REFERENCES

1. S. Araya, X. Cui, N. Li, V. Liso, and S. L, Sahlin, "Power-to-X Technology overview, possibilities and challenges," 2022. https://vbn.aau.dk/ws/portalfiles/portal/514146100/PtX_Report.pdf
2. A. Stirling, "Power-to-X technologies, e-fuels and their applications," *Pinsent Masons*, 2023. https://www.pinsentmasons.com/out-law/guides/power-to-x-technologies-e-fuels-and-their-applications
3. Infinity PV, "What is power-to-X? A complete guide to renewable energy storage and conversion," *Infinity PV.* https://www.infinitypv.com/roll-to-roll-academy/what-is-power-to-x-a-complete-guide-to-renewable-energy-storage-and-conversion
4. Eurowind Energy, "Power-to-X (PtX): an enabler of the future," 2025. https://eurowindenergy.com/power-to-x-ptx
5. M. Genovese, A. Schlüter, E. Scionti, F. Piraino, O. Corigliano, and P. Fragiacomo, "Power-to-hydrogen and hydrogen-to-X energy systems for the industry of the future in Europe," *Int. J. Hydrogen Energy*, vol. 48, no. 44, pp. 16545–16568, 2023. doi: 10.1016/j.ijhydene.2023.01.194
6. A. M. Sadeq et al., "Hydrogen energy systems: technologies, trends, and future prospects," *Sci. Total Environ.*, vol. 939, p. 173622, Aug. 2024. doi: 10.1016/j.scitotenv.2024.173622
7. M. M. Hossain Bhuiyan and Z. Siddique, "Hydrogen as an alternative fuel: a comprehensive review of challenges and opportunities in production, storage, and transportation," *Int. J. Hydrogen Energy*, vol. 102, pp. 1026–1044, Feb. 2025. doi: 10.1016/j.ijhydene.2025.01.033
8. J. Gómez and R. Castro, "Green hydrogen energy systems: a review on their contribution to a renewable energy system," *Energies*, vol. 17, no. 13, pp. 1–41, 2024. doi: 10.3390/en17133110
9. R. Jain, N. L. Panwar, Chitranjan Agarwal, and T. Guta, A comprehensive review on unleashing the power of hydrogen: revolutionizing energy systems for a sustainable future, no. May. 2024. doi: 10.1007/s11356-024-33541-1
10. A. Marghani, E. A. Chater, I. Bouganssa, N. Sefiani, A. Elakkary, and I. Lahlouh, "Hydrogen production and applications: a review," *E3S Web Conf.*, vol. 469, pp. 1–7, 2023. doi: 10.1051/e3sconf/202346900088
11. P. Kumar and L. Fiori, "Thermochemical and biological routes for biohydrogen production: a review," *Energy Convers. Manag. X*, vol. 23, p. 100659, Jul. 2024. doi: 10.1016/j.ecmx.2024.100659
12. N. K. Obiora, C. O. Ujah, C. O. Asadu, F. O. Kolawole, and B. N. Ekwueme, "Production of hydrogen energy from biomass: prospects and challenges," *Green Technol. Sustain.*, vol. 2, no. 3, p. 100100, Sep. 2024. doi: 10.1016/j.grets.2024.100100
13. M. Yue et al., "Hydrogen energy systems : a critical review of technologies, applications, trends and challenges," *Renew. Sustain. Energy Rev.,*vol. 146, pp. 0–34, 2023.
14. A. M. Sadeq, "Advances and challenges in hydrogen energy : a review," *Chem. Eng. Process Tech.*, vol. 9, pp. 1–5, 2024.
15. R. Tichler, S. Bauer, and H. Böhm, "Power-to-gas," in *Storing Energy*, Elsevier, 2022, pp. 595–612. doi: 10.1016/B978-0-12-824510-1.00010-6
16. S. Schiebahn, T. Grube, M. Robinius, V. Tietze, B. Kumar, and D. Stolten, "Power to gas: technological overview, systems analysis and economic assessment for a case study in Germany," *Int. J. Hydrogen Energy*, vol. 40, no. 12, pp. 4285–4294, Apr. 2015. doi: 10.1016/j.ijhydene.2015.01.123
17. C. Wulf, J. Linssen, and P. Zapp, "Power-to-gas – concepts, demonstration, and prospects," in *Hydrogen Supply Chains*, Elsevier, 2018, pp. 309–345. doi: 10.1016/B978-0-12-811197-0.00009-9

18. S. Drünert, U. Neuling, T. Zitscher, and M. Kaltschmitt, "Power-to-liquid fuels for aviation – processes, resources and supply potential under German conditions," *Appl. Energy*, vol. 277, p. 115578, Nov. 2020. doi: 10.1016/j.apenergy.2020.115578
19. W. Suwaileh et al., "Exploring hydrogen fuel as a sustainable solution for zero-emission aviation: production, storage, and engine adaptation challenges," *Int. J. Hydrogen Energy*, vol. 121, pp. 304–325, Apr. 2025. doi: 10.1016/j.ijhydene.2025.03.348
20. J. G. Speight, "Hydrocarbons from synthesis gas," in *Handbook of Industrial Hydrocarbon Processes*, Elsevier, 2020, pp. 343–386. doi: 10.1016/B978-0-12-809923-0.00008-4
21. International Civil Aviation Organisation, "Power-to-Liquids (PTL) sustainable fuels for aviation," 2015.
22. P. Levi and A. F. Pales, "From energy to chemicals," *IEA*, 2018.
23. M. de Graaff and M. Goswami, "Power-2-chemicals," Volta Chem.
24. V. Sakthimurugan, G. Lakshmikanth, N. Balaji, R. Roopashree, D. Kumar, and Y. Devarajan, "Green hydrogen revolution: advancing electrolysis, market integration, and sustainable energy transitions towards a net-zero future," *Results Eng.*, vol. 26, p. 104849, Jun. 2025. doi: 10.1016/j.rineng.2025.104849
25. K. Sanford, G. Chotani, N. Danielson, and J. A. Zahn, "Scaling up of renewable chemicals," *Curr. Opin. Biotechnol.*, vol. 38, pp. 112–122, Apr. 2016. doi: 10.1016/j.copbio.2016.01.008
26. ENERGYNEST, "Power-to-heat: industrial utilisation of heat storage systems," *Insights*, 2024. https://energy-nest.com/power-to-heat-industrial-utilisation/
27. BMWK, "What exactly is meant by 'power-to-heat''?,*Energiewende Direkt*, 2016. https://www.bmwk-energiewende.de/EWD/Redaktion/EN/Newsletter/2016/07/Meldung/direkt-answers.html
28. European Energy, "Power-to-X substitutes fossil-fuels and decarbonises industries," 2025. https://europeanenergy.com/green-solutions/ptx/
29. SolarWatt, "Sector coupling: key technologies of the energy transition," 2025. https://www.solarwatt.co.uk/knowledge-base/sector-coupling
30. K.-T. Lee, Y.-S. Cai, Q.-Y. Hou, T.-J. Hsu, and W.-H. Hsu, "A brief overview of green hydrogen on production, regulations, and commercialization," *Green Energy Fuel Res.*, Aug. 2024. doi: 10.53941/gefr.2024.100002
31. I. Zaiter, M. Ramadan, A. Bouabid, A. Mayyas, M. El-Fadel, and T. Mezher, "Enabling industrial decarbonization: framework for hydrogen integration in the industrial energy systems," *Renew. Sustain. Energy Rev.*, vol. 203, p. 114782, Oct. 2024. doi: 10.1016/j.rser.2024.114782
32. Ministry of Foreign Affairs of Denmark, "Denmark announces new power-to-X strategy." https://investindk.com/insights/denmark-announces-new-power-to-x-strategy
33. M. Batool, O. Sanumi, and J. Jankovic, "Application of artificial intelligence in the materials science, with a special focus on fuel cells and electrolyzers," *Energy AI*, vol. 18, p. 100424, Dec. 2024. doi: 10.1016/j.egyai.2024.100424
34. S. M. Shojaei, R. Aghamolaei, and M. R. Ghaani, "Recent advancements in applying machine learning in power-to-X processes: a literature review," *Sustainability*, vol. 16, no. 21, p. 9555, Nov. 2024. doi: 10.3390/su16219555
35. D. D. K. Wayo, L. Goliatt, and D. Ganji, "AI and quantum computing in binary photocatalytic hydrogen production," *arxiv*, 2024. https://arxiv.org/abs/2501.00575
36. Y. Gu et al., "Design and application of electrocatalyst based on machine learning," *Interdiscip. Mater.*, vol. 4, no. 3, pp. 456–479, May 2025. doi: 10.1002/idm2.12249
37. D. Khalafallah et al., "Machine learning-driven breakthroughs in water electrolysis and supercapacitors," *Mater. Chem. Front.*, vol. 9, no. 15, pp. 2322–2353, 2025. doi: 10.1039/D5QM00326A
38. P. Tiwari, "Digital twins & IoT in metal fabrication: optimizing production process," *METALBOOK*,2025.https://www.metalbook.com/blogs/digital-twins-and-iot-in-metal-fabrication-optimizing-production-process/

39. OneMind NG, "Smart industry meets digital twins," *One Mind: New Generation*, 2025.
40. O. H. Laguna, P. F. Lietor, F. J. I. Godino, and F. A. Corpas-Iglesias, "A review on additive manufacturing and materials for catalytic applications: milestones, key concepts, advances and perspectives," *Mater. Des.*, vol. 208, p. 109927, Oct. 2021. doi: 10.1016/j.matdes.2021.109927
41. A. Soliman, N. AlAmoodi, G. N. Karanikolos, C. C. Doumanidis, and K. Polychronopoulou, "A review on New 3-D printed materials' geometries for catalysis and adsorption: paradigms from reforming reactions and CO_2 capture," *Nanomaterials*, vol. 10, no. 11, p. 2198, Nov. 2020. doi: 10.3390/nano10112198
42. I. Rezić Meštrović and M. Somogyi Škoc, "Nanofiber-based innovations in energy storage systems," *Polymers*, vol. 17, no. 11, p. 1456, May 2025. doi: 10.3390/polym17111456
43. NANOGRAFI, "FUEL CELL DEVELOPMENT WITH NANOtechnology," *NANOGRAFI.* https://shop.nanografi.com/blog/fuel-cells-development-with-nanotechnology-728a03/#:~:text=Various types of carbon nanomaterials,platinum in oxygen reduction reactions.
44. M. Z. Ansari, S. N. Banitaba, S. Khademolqorani, I. Kamika, and V. V. Jadhav, "Overlooked promising green features of electrospun cellulose-based fibers in lithium-ion batteries," *ACS Omega*, vol. 8, no. 46, pp. 43388–43407, Nov. 2023. doi: 10.1021/acsomega.3c05068
45. M. Rahman et al., "A review on nanomaterial-based additive manufacturing: dynamics in properties, prospects, and challenges," *Prog. Addit. Manuf.*, vol. 9, no. 4, pp. 1197–1224, Aug. 2024. doi: 10.1007/s40964-023-00514-8
46. N. F. Raduwan, N. Shaari, S. K. Kamarudin, M. S. Masdar, and R. M. Yunus, "An overview of nanomaterials in fuel cells: synthesis method and application," *Int. J. Hydrogen Energy*, vol. 47, no. 42, pp. 18468–18495, May 2022. doi: 10.1016/j.ijhydene.2022.03.035
47. O. Cardozo et al., "Zinc oxide nanostructures for third generation solar cells: a comprehensive review," *Sol. Energy*, vol. 299, p. 113710, Oct. 2025. doi: 10.1016/j.solener.2025.113710
48. L. Kreiling, D. K. R. Robinson, and D. Winickoff, "Collaborative platforms for innovation in advanced materials," *OECD Sci. Technol. Ind. Policy Pap.*, no. 95, p. 67, 2020. https://www.proquest.com/working-papers/collaborative-platforms-innovation-advanced/docview/2805208540/se-2?accountid=6802%0Ahttps://media.proquest.com/media/hms/PFT/1/JCEOR?_a=ChgyMDI1MDUyNzAyMTIyMjU4MToyMTEzMzgSBjExNTY0MxoKT05FX1NFQVJDSCINMTQ3LjQ3LjM5Lj
49. D. Winickoffi, L. Kreilingi, M. Borowieckii, H. Gardeni, and J. Philpi, "Collaborative platforms for emerging technology: creating convergence spaces," *OECD Sci. Technol. Ind. Policy Pap.*, no. 109, p. 85, 2021. https://www.oecd-ilibrary.org/science-and-technology/collaborative-platforms-for-emerging-technology_ed1e030d-en
50. L. Talirz et al., "Materials cloud, a platform for open computational science," *Sci. Data*, vol. 7, no. 1, p. 299, Sep. 2020. doi: 10.1038/s41597-020-00637-5
51. A. K. Sahu et al., "Material needs for power-to-X systems for CO_2 utilization require a life cycle approach," *Chem. Sci.*, vol. 16, no. 14, pp. 5819–5835, 2025. doi: 10.1039/D4SC07752K
52. E. E. Nzaba Madila, A. Makhsoos, M. M. Shanbhag, and B. G. Pollet, "Advancements in electrolyser stack performance: a comprehensive review of Latest technologies and efficiency strategies," *Int. J. Hydrogen Energy*, vol. 144, pp. 1168–1189, Jul. 2025. doi: 10.1016/j.ijhydene.2025.04.047
53. X. Gao et al., "Next-generation green hydrogen: progress and perspective from electricity, catalyst to electrolyte in electrocatalytic water splitting," *Nano-Micro Lett.*, vol. 16, no. 1, p. 237, Dec. 2024. doi: 10.1007/s40820-024-01424-2
54. C. Zhan et al., "Key intermediates and Cu active sites for CO_2 electroreduction to ethylene and ethanol," *Nat. Energy*, vol. 9, no. 12, pp. 1485–1496, Sep. 2024. doi: 10.1038/s41560-024-01633-4

55. L. Xie, Y. Jiang, W. Zhu, S. Ding, Y. Zhou, and J.-J. Zhu, "Cu-based catalyst designs in CO_2 electroreduction: precise modulation of reaction intermediates for high-value chemical generation," *Chem. Sci.*, vol. 14, no. 47, pp. 13629–13660, 2023. doi: 10.1039/D3SC04353C
56. Y. Liu et al., "Electrochemical reduction of carbon dioxide to ethanol: reaction conditions and tuning strategies," *Coord. Chem. Rev.*, vol. 544, p. 216981, Dec. 2025. doi: 10.1016/j.ccr.2025.216981
57. C. Chen et al., "Harnessing single-atom catalysts for CO_2 electroreduction: a review of recent advances," *EES Catal.*, vol. 2, no. 1, pp. 71–93, 2024. doi: 10.1039/D3EY00150D
58. Y. Ji, J. Du, A. Chen, X. Gao, and M. Peng, "Transition metal-nitrogen-carbon single-atom catalysts enhanced CO_2 electroreduction reaction: a review," *ChemSusChem*, vol. 18, no. 3, Feb. 2025. doi: 10.1002/cssc.202401557
59. L. Jiao et al., "Insight into the metal-free electrocatalysis of heteroatom-doped carbon nanocages in competitive CO_2 reduction and H_2 evolution," *Nano Res.*, vol. 18, no. 2, p. 94907171, Feb. 2025. doi: 10.26599/NR.2025.94907171
60. Y. Sun, K. Zhao, X. Deng, M. Zhang, X. Wang, and W. Wang, "Metal-free Se-based tetra-doped carbon catalyst for high-selective electro-reduction of CO_2 into CO," *J. Environ. Chem. Eng.*, vol. 11, no. 5, p. 110435, Oct. 2023. doi: 10.1016/j.jece.2023.110435
61. D. Luo et al., "Cu-based bimetallic catalysts for electrochemical CO_2 reduction: before and beyond the tandem effect," *Nanoscale*, vol. 17, no. 15, pp. 9057–9071, 2025. doi: 10.1039/D4NR04790G
62. Q. Wang, H. Wei, P. Liu, Z. Su, and X.-Q. Gong, "Recent advances in copper-based catalysts for electrocatalytic CO_2 reduction toward multi-carbon products," *Nano Res. Energy*, vol. 3, no. 3, p. e9120112, Sep. 2024. doi: 10.26599/NRE.2024.9120112
63. X. Meng et al., "Cu/Zn bimetallic catalysts prepared by facial potential steps electrodeposition favoring Zn deposition and grain boundary formation for efficient CO_2ER to ethylene," *Fuel*, vol. 369, p. 131775, Aug. 2024. doi: 10.1016/j.fuel.2024.131775
64. K. Khoiruddin, C. S. A. Nandar, S. Kawi, T. M. Lim, and I. G. Wenten, "Advanced proton exchange membranes for high-efficiency fuel cells: material innovations and durability optimization," *Energy Mater.*, vol. 5, no. 11, Aug. 2025. doi: 10.20517/energymater.2025.62
65. Z. Zhang, Y. Dou, W. Zhang, L. Xu, and Y. Wang, "A fiberglass-cloth-reinforced perfluorosulfonic acid membrane," *Membranes*, vol. 15, no. 6, p. 166, Jun. 2025. doi: 10.3390/membranes15060166
66. J. Li, Q. Cai, and B. Amini Horri, "Highly conductive and stable electrolytes for solid oxide electrolysis and fuel cells: fabrication, characterisation, recent progress and challenges," *Mater. Adv.*, vol. 6, no. 1, pp. 39–83, 2025. doi: 10.1039/D4MA00690A
67. E. V. Gordeev and D. A. Osinkin, "Step-by-step strategy to improve the performance of the $(La,Sr)(Ga,Mg)O_3$-δ electrolyte for symmetrical solid oxide fuel cells," *Ceram. Int.*, vol. 50, no. 22, pp. 47395–47401, Nov. 2024. doi: 10.1016/j.ceramint.2024.09.089
68. R. Mokete, F. Mikšík, R. Selyanchyn, N. Takata, K. Thu, and T. Miyazaki, "Fabrication of novel mixed matrix polymer electrolyte membranes (PEMs) intended for renewable hydrogen production via electrolysis application," *Energy Adv.*, vol. 3, no. 5, pp. 1019–1036, 2024. doi: 10.1039/D3YA00503H
69. Wikipedia, "Solid oxide electrolyzer cell," *Wikipedia*, 2025. https://en.wikipedia.org/wiki/Solid_oxide_electrolyzer_cell?utm_source=chatgpt.com
70. A. Sciazko, T. Shimura, Y. Komatsu, and N. Shikazono, "Ni-GDC and Ni-YSZ electrodes operated in solid oxide electrolysis and fuel cell modes," *J. Therm. Sci. Technol.*, vol. 16, no. 1, p. JTST0013, 2021. doi: 10.1299/jtst.2021jtst0013
71. R. Daiyan, I. MacGill, and R. Amal, "Opportunities and challenges for renewable power-to-X," *ACS Energy Lett.*, vol. 5, no. 12, pp. 3843–3847, Dec. 2020. doi: 10.1021/acsenergylett.0c02249

72. W. Mabhulusa, K. E. Sekhosana, and X. Fuku, "The impact and performance of carbon-supported platinum group metal electrocatalysts for fuel cells," *Int. J. Electrochem. Sci.*, vol. 19, no. 4, p. 100524, Apr. 2024. doi: 10.1016/j.ijoes.2024.100524
73. UNSW, "Highly efficient and low cost photovoltaic-electrolysis (PVE) system to generate hydrogen by harvesting the full spectrum of sunlight." https://www.pcrg.unsw.edu.au/research/featured-projects/highly-efficient-and-low-cost-photovoltaic-electrolysis-pve-system#:~:text=Project summary,high cost and modest efficiency.
74. M. V. Bram, J. Liniger, S. S. Majidabad, H. R. Shabani, M. P. R. Teles, and X. Cui, "Challenges in Power-to-X: a perspective of the configuration and control process for E-methanol production," *Int. J. Hydrogen Energy*, vol. 76, pp. 315–325, 2024. doi: 10.1016/j.ijhydene.2024.05.273
75. A. Skaaret, "Power-to-X: challenges and possibilities in project development," *Southern Lights*, 2024. https://www.southernlights.io/industry/what-is-power-to-x/
76. International Ptx Hub, "A first look at water demand for green hydrogen and concerns and opportunities with desalination," *International Ptx Hub*, 2024. https://ptx-hub.org/a-first-look-at-water-demand-for-green-hydrogen-and-concerns-and-opportunities-with-desalination/
77. B. Decourt, "Weaknesses and drivers for power-to-X diffusion in Europe. insights from technological innovation system analysis," *Int. J. Hydrogen Energy*, vol. 44, no. 33, pp. 17411–17430, Jul. 2019. doi: 10.1016/j.ijhydene.2019.05.149
78. C. Wulf, P. Zapp, and A. Schreiber, "Review of power-to-X demonstration projects in Europe," *Front. Energy Res.*, vol. 8, no. September, pp. 1–12, 2020. doi: 10.3389/fenrg.2020.00191
79. G. M. E. (GME), "Top 10 green hydrogen producing countries," 2024. https://www.globalmarketestimates.com/blogpost/top-10-green-hydrogen-producing-countries-193
80. PtX Development, "PtX Development and KGAL sign contract for large-scale hydrogen project in Germany." https://ptx-development.de/en/newsroom/
81. Uniper, "Green Wilhelmshaven: to new horizons." https://www.uniper.energy/solutions/energy-transformation-hubs/energy-transformation-hub-northwest/green-wilhelmshaven
82. European Commission, "HydrOxy: HydrOxy Hub Walsum," *Hydrogen Innovation Fund*, 2025. https://ec.europa.eu/assets/cinea/project_fiches/innovation_fund/101133035.pdf
83. KfW Development Bank, "Integrated financing for green hydrogen - a perfect fit and from a single source," 2025. https://www.kfw-entwicklungsbank.de/Our-topics/PtX/
84. International Ptx Hub, "Hydrogen Factsheet Japan," 2021. https://ptx-hub.org/publication/hydrogen-factsheet-japan/
85. Fuel Cell Works, "INPEX begins commissioning at Japan's blue hydrogen and ammonia project,"2025. https://fuelcellsworks.com/2025/06/06/h2/inpex-begins-commissioning-at-japan-s-blue-hydrogen-and-ammonia-project
86. P. Martin, "Japanese energy companies launch the country's largest green hydrogen project," *Hydrogeninsight*, 2024. https://www.hydrogeninsight.com/production/japanese-energy-companies-launch-the-countrys-largest-green-hydrogen-project/2-1-1602552?zephr_sso_ott=CxNwxR
87. Department of Climate Change, Energy, The Environment and Water, Australian Government, "Growing Australia's hydrogen industry," *Energy*. https://www.dcceew.gov.au/energy/hydrogen
88. Dansk Industri, "Green Hydrogen/ Ptx market in Australia," 2022. https://www.danskindustri.dk/arrangementer/soeg/arrangementer/internationalt---oceanien/ptx-australia/#:~:text=Australia is world leading within the development,to Asia and other energy importing regions.&text=Key learnings from the market indicate that,h
89. Innovation News Network, "US clean hydrogen technology boosted with $750m investment," 2024. https://www.innovationnewsnetwork.com/us-clean-hydrogen-technology-boosted-with-750m-investment/45283/

90. reNews, "European Energy signs port lease for US PtX plant," *reNews*. https://renews.biz/90808/european-energy-signs-port-lease-for-us-ptx-plant/
91. M. Markosyan, "MAN energy gets job on StormFisher's 200-MW PtX project in N America," *Renewables Now*, 2025. https://renewablesnow.com/news/man-energy-gets-job-on-stormfishers-200-mw-ptx-project-in-n-america-1271872/
92. reNews, "Nova acquires 1GW Texas wind-solar-PtX project," 2024. https://renews.biz/92568/nova-acquires-1gw-texas-wind-solar-ptx-project/#:~:text=Calendar of Events * Connecting Projects to,WIND 2025 16–19 September 2025.
93. GreenGo Energy, "No title," 2025. https://www.greengoenergy.com/news/gge-us-expansion-250212#:~:text=GreenGo Energy has more than 35 GW,chip investors in the renewable energy space.
94. Hydrogen Industry Leaders, "Green ammonia PtX plant to be developed in China," 2025. https://hydrogenindustryleaders.com/green-ammonia-ptx-plant-to-be-developed-in-china/
95. Xinhua, "China's largest offshore solar-hydrogen farm starts operation," *Global Times*, 2025. https://www.globaltimes.cn/page/202501/1326269.shtml
96. Fuel Cell Works, "China begins construction of landmark $290M green hydrogen project in Ningxia Region," 2024. https://fuelcellsworks.com/2024/11/11/green-hydrogen/china-begins-construction-of-landmark-290m-green-hydrogen-project-in-ningxia-region
97. CHN Energy, "China's first marine hydrogen-ammonia-methanol Project Begins Testing," 2025. https://www.ceic.com/gjnyjtwwEn/xwzx/202504/b4186a4d41c6445f85d860e65c9987fd.shtml
98. J. Atchison, "Korea southern power selected as final bidder in national clean power auction," *Ammonia Energy Association*, 2024. https://ammoniaenergy.org/articles/korea-southern-power-selected-as-final-bidder-in-national-clean-power-auction/
99. A. Memija, "Offshore Wind in First Half of 2025," *Offshore Wind*, 2025. https://www.offshorewind.biz/2025/05/27/south-korea-to-tender-1-25-gw-fixed-bottom-offshore-wind-in-first-half-of-2025/
100. Fuel Cells Works, "South Korea to construct world's largest hydrogen fuel cell power plant in Gyeongju," 2024. https://fuelcellsworks.com/2024/11/28/fuel-cells/south-korea-to-construct-world-s-largest-hydrogen-fuel-cell-power-plant-in-gyeongju
101. L. Collins, "South Korea's first nuclear hydrogen project to be built by 2027 by eight partners, including Hyundai and Samsung," *Hydrogen Insight*, 2024. https://www.hydrogeninsight.com/innovation/south-korea-s-first-nuclear-hydrogen-project-to-be-built-by-2027-by-eight-partners-including-hyundai-and-samsung/2-1-1663769
102. The Maritime Executive, "South Korea Funds Project to build world's largest liquid hydrogen carrier," 2025. https://maritime-executive.com/article/south-korea-funds-project-to-build-world-s-largest-liquid-hydrogen-carrier#:~:text=South Korea Funds Project to Build World%27s Largest Liquid Hydrogen Carrier.
103. Green Hydrogen Organisation, "France." https://gh2.org/countries/france
104. HY2 GEN, "France: Sunrhyse." https://www.hy2gen.com/sunrhyse
105. Lhyfe, "Lhyfe announces the construction of a 5th green hydrogen production site in France, the 1st in northern France," 2025. https://www.lhyfe.com/press/lhyfe-announces-the-construction-of-a-5th-green-hydrogen-production-site-in-france-the-1st-in-northern-france/#:~:text=Lhyfe announces the construction of a 5th,nearing completion%2C with commissioning scheduled for 2026.
106. Rystad Energy, "Spain sets sights on dominating regional hydrogen market, on track to hit 2030 national target," 2025. https://www.rystadenergy.com/news/spain-dominating-regional-hydrogen-market
107. C. Currie, "Spain confirms €1.22bn for 2.3GW of green hydrogen valley projects," *H2View*, 2025. https://www.h2-view.com/story/spain-confirms-e1-22bn-for-2-3gw-of-green-hydrogen-valley-projects/2124095.article/

108. International Climate Initiative (IKI), "Enabling long term defossilisation pathways through Power-to-X (PtX Pathways)," 2025. https://www.international-climate-initiative.com/en/project/enabling-long-term-defossilisation-pathways-through-power-to-x-ptx-pathways-20-i-389-global-g-ptx-pathways/#:~:text=As part of the International PtX Hub%2C,of the countries and their individual m
109. International PtX Hub, "Study on Power-to-X production with geothermal energy in El Salvador, Chile, and Kenya," 2023. https://ptx-hub.org/publication/a-geothermal-approach-to-power-to-x/#:~:text=Country(s): Chile%2C El Salvador%2C Kenya.
110. N. H. Prevljak, "World's 'largest' green hydrogen plant construction reaches 80% completion," *Offshore Energy*, 2025. https://www.offshore-energy.biz/worlds-largest-green-hydrogen-plant-construction-reaches-80-completion/#:~:text=NEOM Green Hydrogen Project in,gases company Air Products said.
111. EU, "Winners of first EU-wide renewable hydrogen auction sign grant agreements, paving the way for new European production," Energy, Climate Change, Environment: *Climate Action*, 2024. https://climate.ec.europa.eu/news-your-voice/news/winners-first-eu-wide-renewable-hydrogen-auction-sign-grant-agreements-paving-way-new-european-2024-10-07_en
112. J. Atchison, "Denmark approves national PtX strategy," 2022. https://ammoniaenergy.org/articles/denmark-approves-national-ptx-strategy/#:~:text=Denmark joins India and South Africa in,roadmaps for hydrogen & ammonia production scale-up.
113. GIZ, "Promoting climate-neutral fuels and chemicals." https://www.giz.de/en/worldwide/129317.html
114. Thyssenkrupp Nucera, "Power-to-X (PtX): transforming energy for a sustainable future," *New Era Insights*, 2024. https://www.new-era-insights.com/article/power-to-x-ptx-transforming-energy-for-a-sustainable-future/
115. Eurowind Energy, "Power-to-X (PtX): an enabler of the future," 2025. https://eurowindenergy.com/power-to-x-ptx#:~:text=Our journey towards a greener future can,sectors such as electricity%2C gas and heating.&text=The project is planned to be finished at the end of 2024.
116. Ramboll, "Power-to-X: paving the way for a greener future," 2025. https://www.ramboll.com/net-zero-explorers/power-to-x-explained
117. R. June, "The Catalyst Group Resources™ Power-to-X : techno-economic, commercial and strategic developments for production of energy carrier chemicals, petrochemicals and sustainable fuels Power-to-X : techno-economic, commercial and strategic developments," 2021. https://www.catalystgrp.com/wp-content/uploads/2024/12/Power-to-X-Techno-Economic-Commercial-and-Strategic-Developments.pdf
118. T. Uhorakeye, A. Kopp-Moini, and B. Cohen, "PtX business opportunities in South Africa: renewable hydrogen market potential and value chain analysis," 2024. https://ptx-hub.org/wp-content/uploads/2024/05/Green-Hydrogen-South-Africa_2311_Business-Opportunities-South-Africa-GH2.pdf
119. J. Liu, L. Guo, Z. Dong, Q. Zhao, and L. Li, "PTX: the bridge to a more flexible energy system," in *Proceedings of the 1st International Joint Conference on Energy and Environmental Engineering*, 2021, pp. 22–30. doi: 10.5220/0011104900003355
120. G. Müller, F. Kullmann, J. Linssen, and D. Stolten, "The costs of future energy technologies: a comprehensive review of power-to-X processes," *J. CO_2 Util.*, vol. 92, no. July 2024, 2025. doi: 10.1016/j.jcou.2025.103019
121. Netscribes, "AI meets cloud: A synergy shaping the future of IoT," 2024. https://www.netscribes.com/iot-digital-transformation/
122. M. Pietrasik, A. Wilbik, and P. Grefen, "The enabling technologies for digitalization in the chemical process industry," *Digit. Chem. Eng.*, vol. 12, p. 100161, Sep. 2024. doi: 10.1016/j.dche.2024.100161

9 Digital Solutions for Net-Zero Applications

Optimizing Material Selection and Processes

Suneela Sardar, Aliyu M. Aliyu, Shadab Ahmad, and Faiz Iqbal

9.1 INTRODUCTION

In the contemporary manufacturing landscape, achieving net-zero emissions has emerged as a pivotal objective in response to escalating environmental concerns and stringent regulatory requirements. However, micro-, small-, and medium-sized enterprises (MSMEs) have been lagging in adoption of advanced digital and Industry 4.0 technologies due to the substantial costs and perceived complexities involved. This hesitancy is particularly pronounced among MSMEs, which typically operate with limited financial resources and lack the technical expertise to integrate these innovations seamlessly into their production processes. A considerable number of studies have highlighted the barriers MSMEs face in the adoption of digital technologies. Cost is consistently identified as a primary obstacle. For instance, Müller et al. [1] underscore that high initial investment requirements deter MSMEs from embracing Industry 4.0 technologies. Similarly, a study by Mittal et al. [2] indicated that the financial burden associated with digital transformation is disproportionately high for smaller enterprises, which often results in a reluctance to transition from traditional manufacturing practices. Iqbal et al. [3] briefly described how MSMEs were facing difficulties during COVID-19 due to lack of adoption of digital technologies. Moreover, the lack of technical knowledge and skills exacerbates the situation. Research by Müller and Voigt [4] reveals that MSMEs frequently struggle with the complexity of digital technologies and lack access to skilled labour capable of managing and maintaining such systems. This skill gap further widens the divide between large enterprises and MSMEs in terms of technological adoption and digital innovation.

Despite these challenges, the potential benefits of digital technologies in achieving net-zero goals are significant. Digital automation and other advanced technologies can enhance energy efficiency, reduce waste, and optimize resource use. A study by Kang et al. [5] illustrated that digital technologies can lead to substantial reductions in energy consumption and greenhouse gas emissions in manufacturing processes. Furthermore, implementing Internet of Things (IoT) solutions can facilitate real-time

DOI: 10.1201/9781003582304-12

monitoring and management of energy usage, thereby aligning manufacturing operations with sustainability targets [6]. To bridge the gap, several strategies can be employed to support MSMEs in adopting digital technologies. Financial incentives, such as grants and subsidies, can alleviate the initial investment burden.

Training programmes and collaborative initiatives between industry and educational institutions can help address the skills gap, providing MSMEs with the necessary expertise to implement and manage digital systems. Additionally, the development of user-friendly and scalable digital solutions tailored to the specific needs of MSMEs can facilitate smoother integration and adoption. While the road to digital transformation presents significant challenges for MSMEs, the potential rewards in terms of sustainability and efficiency are substantial. By addressing financial and technical barriers, and fostering an environment conducive to digital innovation, MSMEs can play a crucial role in advancing net-zero goals within the manufacturing sector. This chapter proposes a concept of a digital net-zero awareness system for MSMEs in manufacturing, examining the obstacles they face and proposing actionable solutions to enhance their digital integration and environmental sustainability.

9.2 SUSTAINABLE DEVELOPMENT GOALS

The United Nations Sustainable Development Goals (UN SDGs or just SDGs) adopted by the UN Member States in 2015 serve as a blueprint for achieving better and more sustainable practices globally [7]. Within the context of manufacturing systems, several SDGs are relevant in guiding the development of the net-zero-aware concept. Two are most relevant: SDG7 and SDG9. While SDG7 aims to ensure universal access to affordable, reliable, clean, and sustainable energy, SDG9 focusses on industry, innovation, and infrastructure. For manufacturing MSMEs, transitioning to clean energy sources such as wind, solar, and biomass can significantly reduce their carbon footprints. Furthermore, implementing digital systems within manufacturing facilities aligns with the goals of SDG9, as manufacturing MSMEs should monitor emissions with the aim of optimizing their production processes.

This ensures that resilient infrastructure is built that facilitates innovation and promotes sustainable industrialization. Other SDGs relevant to the framework are SDG11, 12, 13, and 17. SDG11 deals with sustainable cities and communities and emphasizes creating inclusive, safe, and resilient cities which manufacturing SMEs can sometimes be part of. The proposed concept can contribute by promoting net-zero practices within SMEs located in urban areas. Furthermore, responsible consumption and production, through the efficient use of resources and reduction of waste, are advocated by SDG12. By integrating net-zero awareness into manufacturing, MSMEs can minimize waste and increase resource efficiency. Climate Action (SDG13) is also key, as it calls for urgent action to combat climate change due to industrial emissions and their impacts which are already being felt with rising temperatures [8,9]. The proposed real-time monitoring within the proposed framework aligns with this SDG, in that it helps SMEs track emissions and make informed decisions as a result.

Finally, SDG17 (Partnerships for the Goals). It emphasizes collaboration among governments, businesses, non-governmental organizations, and civil society. Specifically, the proposed net-zero-aware concept can help build partnerships between MSMEs, experts in industry, regulators, and policymakers. Only when tools such as this are used, can we collectively work towards global net-zero objectives. In summary, it should be noted that the SDGs are interconnected. Achieving one or two of them often contributes to achieving others and integrating net-zero practices into manufacturing systems as outlined by the SDGs not only addresses environmental concerns of society but also ensures the long-term economic sustainability of the businesses of MSMEs.

9.3 DIGITALIZING MSME SUPPLY CHAINS

The digitalization of supply chains in manufacturing MSMEs can significantly enhance their efficiency, responsiveness, and sustainability. Several strategies and technologies can facilitate this transformation, ensuring that MSMEs remain competitive and contribute to net-zero goals. The technologies vary from application capabilities to level of autonomy in them. Some of the techniques are IoT, cloud computing, blockchain, and big data. IoT enables the connection of physical devices to the internet, allowing for real-time data collection and analysis. In supply chains, IoT devices can monitor inventory levels, track shipments, and manage machinery performance helping MSMEs optimize their operations, reduce waste, and improve energy efficiency [10]. Cloud computing offers scalable and cost-effective solutions for data storage, processing, and management. By leveraging cloud-based platforms, MSMEs can access advanced software tools without substantial upfront investment.

Cloud computing supports various supply chain functions, including demand forecasting, order management, and supplier collaboration [11]. Blockchain technology provides a secure and transparent way to record transactions and track assets across the supply chain. For MSMEs, blockchain can enhance traceability, reduce fraud, and improve trust across the supply chain. This ensures all transactions are verifiable and immutable, fostering greater accountability and efficiency [12]. Big data analytics involves the analysis of large volumes of data to uncover patterns, trends, and insights which can enhance decision-making processes, predict demand fluctuations, and optimize logistics. MSMEs can leverage big data analytics to improve their responsiveness to market changes and enhance operational efficiency [13]. To successfully digitalize their supply chains, MSMEs need support through government initiatives and industry bodies in the form of financial incentives, grants, and subsidies. Additionally, creating awareness programmes and offering technical support can help MSMEs navigate the complexities of digital technologies. Collaboration with larger enterprises, research institutions, and technology providers can offer MSMEs access to advanced technologies and expertise.

Joint ventures, partnerships, and industry clusters can facilitate knowledge sharing and resource pooling. Investing in workforce training and development is crucial for the successful implementation of digital technologies. MSMEs can engage in training programmes, workshops, and certifications to equip their employees with the necessary skills to manage and operate digital systems. Instead of a full-scale transformation, MSMEs can adopt a phased approach to digitalization. Starting with pilot projects and gradually scaling up allows these enterprises to manage risks, assess benefits, and make informed decisions.

Some of the important characteristics that influence the μ-SME supply chains are shown in Figure 9.1. These characteristics show the flexibility and rigidity of MSMEs towards digitalization of their supply chain. MSMEs are smaller in size, which gives them more flexibility towards digitalization but provides resource constraint due to the less revenue as production volumes are low. The smaller structure of MSMEs makes them more responsive, and an informal atmosphere allows them to adjust according to market changes but due to the local focus of MSMEs, they have limited market reach and rely on regional markets. Supply chain efficiency of MSMEs depends on the strong collaboration and relationship with stakeholders. Risk tolerance and innovation vary a lot in the MSMEs sector; some of them are risk-averse but others welcome the creative solutions to overcome the resource constraints [14–16].

Figure 9.2 shows some of the opportunities and challenges in achieving the digitalized MSMEs. Digital tools can help MSMEs in attracting new customers through online platforms and e-commerce-based networks. Technological advancement can also pose the risk of customer data security and privacy, proper cybersecurity measures are necessary to overcome this risk. The current challenges are subsiding the opportunities in achieving digitalization of MSMEs, which can be balanced by carrying out digitalization to achieve net-zero goals. Digitalization will help MSMEs in decarbonizing of manufacturing process and controlling/minimizing emissions to achieve net-zero targets [17].

While digitalization has transformed supply chain visibility and coordination, its potential extends far beyond logistics. Emerging digital tools are now reshaping

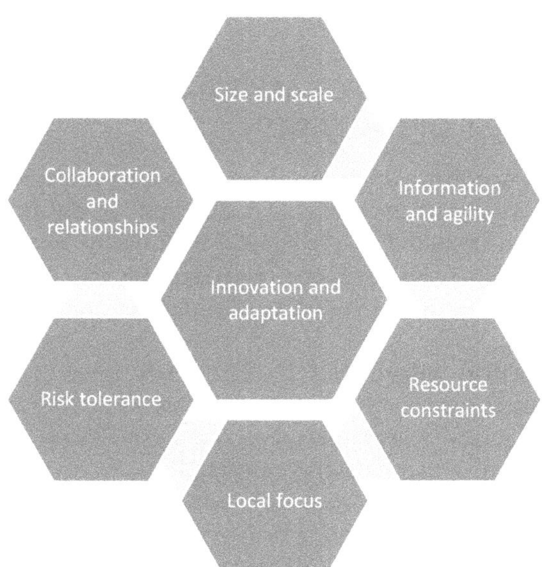

FIGURE 9.1 Distinct characteristic of MSMEs supply chain.

FIGURE 9.2 Opportunities and challenges in achieving digitalized MSMEs.

how materials are chosen, and manufacturing processes are optimized, enabling decisions that simultaneously improve performance, reduce costs, and cut carbon footprints. The next section explores how these innovations can directly support net-zero ambitions.

9.4 DIGITAL SOLUTIONS FOR MATERIAL SELECTION AND PROCESS OPTIMIZATION

Material selection and process optimization are among the most decisive factors in determining a product's environmental footprint. In the past, these choices were often based on limited datasets, supplier familiarity, or trial-and-error prototyping. Today, **digital solutions are transforming this landscape**, enabling evidence-based, real-time decision-making that aligns technical performance with sustainability goals.

AI-driven materials informatics platforms now mine vast, structured and unstructured datasets—drawing from sources such as the Materials Project, MatWeb, and proprietary industry databases—to identify candidate materials that meet precise mechanical, thermal, environmental, and cost criteria. These systems can flag low-carbon alternatives, such as bio-based polymers, recycled alloys, or geopolymer concretes, while predicting their performance under specific operating conditions. For example, a furniture MSME might use such a platform to replace virgin hardwood with engineered bamboo, cutting embodied carbon by 40% without compromising structural integrity.

Integrated life cycle assessment (LCA) tools embedded within CAD or PLM environments allow designers to compare the cradle-to-grave impacts of multiple material and process combinations during the earliest design stages. This real-time feedback loop ensures that sustainability is not an afterthought but a core design parameter. A packaging manufacturer, for instance, could digitally evaluate three biopolymer blends and select the one with the lowest global warming potential and highest recyclability before any physical prototyping begins.

Digital twins—virtual replicas of manufacturing systems—extend this optimization into the production phase. By simulating how different materials behave under varying process parameters such as temperature, pressure, and feed rate, manufacturers can fine-tune operations to minimize energy use, reduce scrap, and shorten commissioning times. An automotive supplier might use a digital twin to optimize injection moulding settings for a recycled composite, achieving a 12% reduction in cycle time and a 15% cut in energy consumption.

Generative design tools, often integrated with additive manufacturing workflows, explore thousands of geometry–material combinations to identify configurations that minimize weight, maximize strength, and reduce waste. A drone manufacturer could apply generative design to create a lightweight frame from recycled aluminium powder, lowering both material use and embodied carbon.

Blockchain-enabled traceability adds a further layer of assurance, recording the provenance, composition, and processing history of materials in a tamper-proof ledger. This not only supports compliance with environmental regulations but also enhances brand credibility. A textile MSME, for example, could certify that its fabrics are sourced entirely from post-consumer waste, meeting ESG reporting requirements and appealing to sustainability-minded customers.

Finally, **cloud-based collaboration platforms** bring together designers, suppliers, and sustainability experts in a shared digital workspace, ensuring that material and process decisions are aligned across the value chain. This reduces delays, improves data accuracy, and embeds sustainability criteria into every stage of product development. A construction SME might use such a platform to coordinate with suppliers on low-carbon concrete mixes, ensuring consistent quality and performance across multiple sites.

For MSMEs, the adoption of these digital solutions offers tangible benefits: reduced prototyping costs, faster innovation cycles, improved compliance with sustainability standards, and a stronger competitive position in markets increasingly shaped by net-zero commitments. In short, **digitalization is no longer optional—it is a strategic enabler of sustainable growth**.

9.5 CARBON FOOTPRINT ANALYSES IN MANUFACTURING

Caron footprint analyses in the manufacturing process revolved around from the initial stage to final product. Manufacturing cycle in MSMEs involves supply chain, production stage, and waste management. Assessing energy consumption and emissions is very important, as they happen at all the stages of the manufacturing process. Carbon footprint analysis of manufacturing process takes into consideration those factors which impact environment including raw materials transportation and processing, production process, distribution of products, and end-of-life treatment

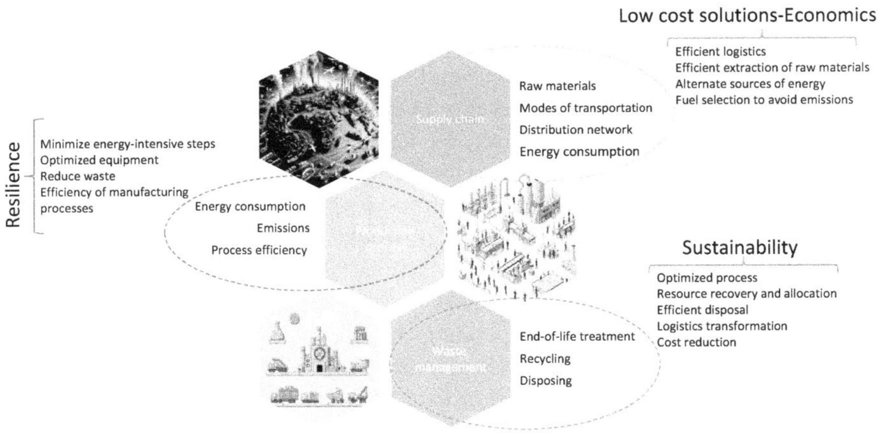

FIGURE 9.3 Achieving net-zero target through digitalization of MSMEs.

of products [18–20]. Net-zero goals mainly focus on resilience, sustainability, and economic benefits. Figure 9.3 shows the contribution of digitalized MSMEs in achieving net-zero targets. Figure 9.4 informs on reducing the carbon footprints at different stages of manufacturing process identified by Figure 9.3. Digitalization of MSMEs will achieve net-zero goals through optimized and efficient processing, alternate and effective sources of energy including renewables, automated supply chain, and efficient waste management system [21,22].

9.6 NET-ZERO AWARENESS

Real-time monitoring of greenhouse gas emissions and providing a database for emission reduction using green practices are the identified needs of the current global net-zero targets. To achieve these targets, it is crucial to build a digitalized real-time system that can predict, prompt, and guide users to achieve the net-zero goals. There is a pressing need for a digital system that can generate and build net-zero awareness in manufacturing MSMEs. The concept of such a system is an interdisciplinary one which must first assess the current practices by industry for emissions, energy efficiency, the barriers, and potential risks. These outcomes must then be fed into the development of the digital platform to act as a net-zero transition system that utilizes I4.0 technologies to provide real-time analysis and prompt net-zero awareness and transformation in industrial organizations primarily MSMEs. The system ought to enable e-connectivity of the supply chain including the physical and computerized aspects and then to integrate net-zero aspects by leveraging current regulations from government websites. Such a system will be able to provide an interactive platform for matchmaking and benchmarking to assess the gaps and enable net-zero transformation.

Globally, the governments, through various initiatives, have urged the private sector and research institutions to develop technologies to drive their respective countries towards becoming a low-carbon economic powerhouse. In the UK, these include via the Hydrogen and Industrial Decarbonization strategies [23]. It must be recognized that data and digitalization are central to these, and thus, the net-zero awareness system merits a development. Critical research is to be carried out in developing such a digital platform that can itself become net-zero aware and then enable the end-users/stakeholders to utilize its services to assess their own situation in achieving net zero.

The development will involve utilizing key I4.0 technologies including machine learning and artificial intelligence (AI) through which the system will be able to first give an overview of current situation of the end-users and then will also be able to prompt and guide them in real-time towards taking necessary steps in achieving net-zero goals. Not only can this be a digital system for an SME or MSME, rather it will be a net-zero-aware digital platform that can ensure the activities of end-users are aligned and contribute towards achieving the net-zero goals by utilizing the global net-zero guidelines. Solutions must be based on the gaps in current practices, i.e., by optimizing energy usage and making use of resources that contribute to minimum emissions. The digital platform will in the longer term make users aware of how net zero will be achieved if green sources were used instead of carbon-emitting alternatives and what would be the optimum way of usage for them.

The digital system must carry out some key activities both before and after its development. User-centred design approach must be followed for the development, the first research activity in which is to carry out a needs assessment i.e., a study of the current practices of how relevant stakeholders aim to achieve net zero and what are the gaps in that context. This can be done through policy dialogues with different stakeholders to understand and communicate existing partnership process/policies. Review of academic literature and sector practice in establishing and managing net-zero emissions practices are required to be done. Such a needs assessment will result in relevant data on current practices and clear gaps which can be drafted into the development of the digital platform for its design and development, keeping gaps in mind and security at the core of the development. Figure 9.4 shows the concept of net-zero-aware digital system applied to a hydrogen production supply chain by a MSME. It must be noted that

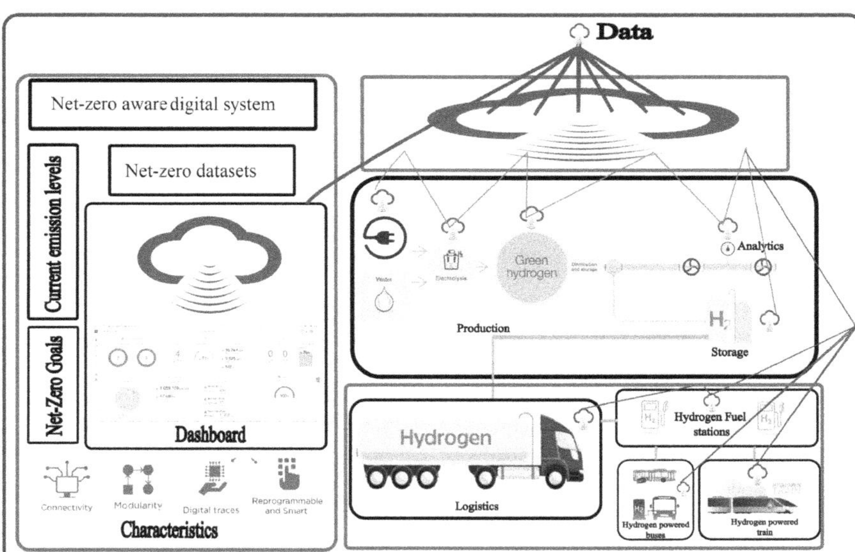

FIGURE 9.4 Concept of the digital net-zero-awareness system.

the application used to describe digital system is for hydrogen production; however, the digital system capabilities described for the concept proposed are equally applicable for any other MSME and hydrogen is only an example.

Integrating material selection into net-zero-awareness platforms ensures that sustainability is embedded not just in supply chains but in the very DNA of product design. By linking AI-driven materials databases and real-time LCA tools to the same digital platform used for emissions tracking, MSMEs can receive ranked recommendations for low-carbon materials based on cost, availability, and performance. These platforms can also simulate process adjustments—via digital twins—before adoption, reducing downtime and risk. Automated alerts flag new sustainable materials or regulatory changes, while blockchain-enabled traceability verifies provenance and recycled content. Embedding these capabilities within a net-zero-awareness system transforms it from a passive reporting tool into an active decision-making engine that drives continuous improvement toward decarbonization.

9.7 CONCLUSION

This chapter has outlined a conceptual aspect to develop a net-zero-awareness building digital system for manufacturing MSMEs. By digitalizing the manufacturing supply chain, these enterprises can significantly enhance their sustainability efforts through real-time monitoring and analysis of emissions. The integration of net-zero aspects within digital systems not only provides SMEs with immediate insights into their carbon footprint but also offers actionable prompts to guide them towards achieving their sustainability goals. The proposed system concept can address the unique challenges faced by MSMEs, such as limited resources and technical expertise, by offering low-cost, scalable solutions tailored to their needs. By leveraging digital technologies, MSMEs can streamline their operations, reduce emissions, and build net-zero awareness both internally and among end-users of their products. This digital acceleration is crucial for aligning manufacturing practices with sustainable development goals, ensuring that even the smallest enterprises contribute effectively to the global effort of mitigating climate change.

REFERENCES

1. Müller JM, Buliga O, Voigt KI. Fortune favors the prepared: how SMEs approach business model innovations in Industry 4.0. *Technological Forecasting and Social Change* 2018; 132: 2–17.
2. Mittal S, Khan MA, Romero D, Wuest T. Smart manufacturing: characteristics, technologies and enabling factors. *Proceedings of the Institution of Mechanical Engineers, Part B: Journal of Engineering Manufacture* 2019; 233(5): 1342–61.
3. Iqbal F, Semwal T, Stokes AA. Cyber-physical systems based smart manufacturing of disinfectants: a need, and solution driven by COVID-19 pandemic. In *International Conference on Flexible Automation and Intelligent Manufacturing*. Cham: Springer Nature Switzerland, 2023, pp. 117–24.
4. Müller JM, Voigt KI. Sustainable industrial value creation in SMEs: a comparison between industry 4.0 and made in China 2025. *International Journal of Precision Engineering and Manufacturing-Green Technology* 2018; 5: 659–70.

5. Kang MJ, Oah S, Chang SS. Exploring the relationship between eco-innovation and corporate sustainability: evidence from Korea. *Environmental Science and Pollution Research* 2020; 27(5): 4866–80.
6. Jabbour CJC, de Sousa Jabbour ABL, Sarkis J, Godinho Filho M. Unlocking the circular economy through new business models based on large-scale data: an integrative framework and research agenda. *Technological Forecasting and Social Change* 2019; 144: 546–52.
7. United Nations. THE 17 GOALS | Sustainable Development. 2015. https://sdgs.un.org/goals
8. Karayil A, Elseragy A, Aliyu, AM. An assessment of CO_2 capture technologies towards global carbon net neutrality. *Energies* 2024; 17(6): 1460.
9. Webber M., Aliyu A, Jin S.-H, Sadiq MM, Sohani, B, Elseragy A. A techno-economic review of direct air capture of moisture processes: sustainable versus energy-intensive methods. *International Journal of Environmental Science and Technology* 2024; 22: 579–613.
10. Jabbour CJC, de Sousa Jabbour ABL, Sarkis J, Godinho Filho M. Unlocking the circular economy through new business models based on large-scale data: an integrative framework and research agenda. *Technological Forecasting and Social Change* 2020; 144: 118–25.
11. Mittal S, Khan MA, Romero D, Wuest T. Smart manufacturing: characteristics, technologies and enabling factors. *Journal of Engineering Manufacture* 2019; 233(5): 1342–61.
12. Kshetri N. 1 Blockchain's roles in meeting key supply chain management objectives. *International Journal of Information Management* 2018; 39: 80–9.
13. Wamba SF, Gunasekaran A, Akter S, Ren, SJF, Dubey R, Childe SJ. Big data analytics and firm performance: effects of dynamic capabilities. *Journal of Business Research* 2017; 70: 356–65.
14. Hong P, Jeong J. Supply chain management practices of SMEs: from a business growth perspective. *Journal of Enterprise Information Management* 2006; 19(3): 292–302.
15. Kerekes V, Felföldi J. Supply chain management practices for SMEs. *Abstract: Applied Studies in Agribusiness and Commerce* 2020; 14: 89–96.
16. OECD. SMEs in more resilient, sustainable and circular supply chains. In Lamia Kamal-Chaoui (Ed.). *OECD SME and Entrepreneurship Outlook*. Paris: OECD Publishing, 2023.
17. Understanding Manufacturing Energy and Carbon Footprints. https://www1.eere.energy.gov/manufacturing/pdfs/understanding_energy_footprints.pdf
18. How to Calculate the Carbon Footprint of Your Products. https://www.rolandberger.com/en/Insights/Publications/How-to-calculate-the-carbon-footprint-of-your-products.html
19. Manufacturing Energy and Carbon Footprint Definitions and Assumptions. https://www.energy.gov/eere/amo/articles/manufacturing-energy-and-carbon-footprint-definitions-and-assumptions-october
20. Manufacturing Energy and Carbon Footprint Assumptions and Definitions. https://www1.eere.energy.gov/manufacturing/pdfs/footprints_assumptions_definitions.pdf
21. Scrucca F, Barberio G, Fantin V, Porta PL, Barbanera M. Carbon footprint: concept, methodology and calculation. Carbon footprint case studies: municipal solid waste management. In Subramanian Senthilkannan Muthu (Ed.). *Sustainable Road Transport and Carbon Sequestration*, Singapore: Springer, 2021, pp. 1–31.
22. Thrassou A, Uzunboylu N, Vrontis D, Christofi M. Digitalization of SMEs: A review of opportunities and challenges. In Alkis Thrassou, Demetris Vrontis, Yaakov Weber, S. M. Riad Shams, Evangelos Tsoukatos (Eds.). *The Changing Role of SMEs in Global Business: Volume II: Contextual Evolution across Markets, Disciplines and Sectors*, Singapore: Springer, 2020, pp. 179–200.
23. Acharya A. Scaling-up green hydrogen development with effective policy interventions. *Journal of Sustainable Development* 2022; 15(5): 135–49.

10 Future Research Direction for Material Innovations in Net-Zero Applications

Faiz Iqbal

This book explores the critical intersection of materials science, digital engineering, and sustainability. It aims to provide researchers, engineers, technologists, and decision-makers with insights into how digital tools are transforming material innovation for net-zero applications across various sectors, including energy storage, lightweight structures, and carbon capture. Through real-world examples, emerging frameworks, and practical insights, the book examines how digital tools are transforming the way we innovate, collaborate, and deliver impact in this crucial field.

This concluding chapter synthesizes the key themes presented throughout the book and outlines promising future research directions in the digitalization of material innovations for net-zero applications.

The chapters in this book collectively highlight the immense potential and necessity of digitalizing material innovations to achieve global net-zero emissions targets. This overarching theme resonates throughout the book's three sections, each contributing a unique perspective to the complex challenge of decarbonization.

Section I laid the foundational understanding by exploring specific material advancements crucial for reducing carbon footprints. **Chapter 2** delved into the specifics of amine-based materials, emphasizing their maturity as a technology for CO_2 capture, their underlying chemistry, and the typical process morphology of such systems. The chapter also discussed strategies for mitigating the degradation of amine-based materials, including the use of hybrid solvents and their innovative applications. Crucially, it highlighted the pivotal role of artificial intelligence (AI) and machine learning (ML) techniques in improving real-world processes and reducing energy penalties associated with CO_2 capture, identifying both challenges and opportunities for widespread deployment. **Chapter 3** then shifted focus to membrane materials, illustrating how digitalization and computational developments are revolutionizing their design and optimization. This includes the application of computational modelling and AI techniques, such as high-throughput screening and insights from molecular modelling, to significantly improve predictive power and efficiency in membrane research. Furthermore, the chapter explored how optimization techniques guarantee the scalability and performance of membrane systems, concluding

DOI: 10.1201/9781003582304-13

by outlining the potential for digitalized membrane technologies to transform the development of high-performance and sustainable membrane applications.

Section II provided a realistic perspective on the adoption and deployment of these crucial innovations. **Chapter 4** addressed the significant role of the built sector in global emissions, emphasizing the need to cut total emissions by 50% before 2030 and fully decarbonize by 2050 to meet Paris Agreement objectives. It highlighted cement manufacturing as a key area for decarbonization efforts due to its high emissions, proposing various levers for cement and concrete manufacturers to decarbonize along the value chain to meet increasing demand for sustainably produced materials. **Chapter 5** expanded on the broader landscape of innovative materials driving progress towards net-zero goals, emphasizing their critical role in achieving a sustainable future. This chapter covered cutting-edge materials essential for carbon capture, low-carbon hydrogen production and storage, and green ammonia production and conversion, all vital for mitigating climate change. **Chapter 6** specifically delved into the practical challenges of implementing next-generation carbon capture materials (NG-CCMs). While these materials perform well in controlled laboratory conditions, their scalability in large industrial set-ups faces significant technical, economic, logistical, and policy/regulatory barriers. The chapter provided a detailed analysis of these barriers, along with potential solutions and real-world case studies from industries such as cement, paint, automotive, and logistics, aiming to bridge the gap between cutting-edge research and industrial application. Finally, **Chapter 7** offered a crucial regional case study, reviewing the development of solar energy systems in Sub-Saharan Africa. Despite the region's vast renewable energy resources, many households remain without electricity. The chapter highlighted currently deployed solar infrastructure, critical constraints hindering widespread adoption, and recommended initiatives and strategic policies to attract international support for sustainable solar projects and climate change mitigation.

Section III brought the core focus back to the digital tools and solutions that enable this profound transformation. **Chapter 8** explored the advent of Power-to-X (PtX) technologies as innovative solutions for integrating variable renewable energy sources into broader energy systems across various sectors. It detailed how PtX encompasses a suite of conversion pathways to transform surplus electricity from renewable sources into hydrogen (Power-to-Hydrogen), synthetic gases and fuels (Power-to-Gas, Power-to-Liquids), chemicals (Power-to-Chemicals), and heat (Power-to-Heat). The chapter emphasized PtX's promising opportunities for long-term energy storage, sector coupling, and decarbonization of hard-to-abate sectors, while also examining real-world deployments, international strategies, and key policy frameworks driving PtX development. It acknowledged significant challenges, including high capital costs, infrastructure limitations, and regulatory gaps that must be addressed for effective scaling. **Chapter 9** proposed a concept for developing a net-zero-aware digital system specifically for manufacturing facilities, with a focus on micro-, small-, and medium-sized enterprises (MSMEs). The chapter explained how digitalizing the manufacturing supply chain by connecting physical aspects can lead to sustainability. It further detailed the integration of net-zero aspects with real-time monitoring and analysis of emissions from production processes in manufacturing MSMEs, thereby providing overall net-zero awareness for both SMEs and their end-users. The suggested concept system aims to offer real-time prompts for SME challenges and provide up-to-date comparisons on their progress towards net-zero goals, demonstrating how low-cost digital solutions can help reduce carbon footprints and achieve sustainability.

Overall, the book underscores that achieving net zero is a multifaceted challenge requiring not only novel material development but also sophisticated digital solutions for monitoring, optimization, and integration across diverse industrial sectors. The convergence of material science and digital engineering is paramount to overcome the existing barriers and accelerate the adoption of sustainable practices.

10.1 FUTURE RESEARCH DIRECTIONS

The comprehensive insights gleaned from this book open several promising avenues for future research and development, building upon the foundations laid by the contributing chapters:

1. **Advanced digital twin integration for material innovation:** While the book touches upon digitalization, there is a significant need for further research into developing more comprehensive, high-fidelity, and interoperable digital twin frameworks. These advanced digital twins should be capable of simulating not only the intrinsic properties and performance of new materials but also their entire life cycle, from the initial stages of raw material extraction and processing, through manufacturing and application, to their end-of-life scenarios, including recycling or disposal. This holistic approach, integrating environmental impact and economic viability into the digital twin, would enable predictive maintenance, significantly optimize resource utilization, and facilitate the proactive identification of sustainability improvements for novel materials, pushing beyond current capabilities to truly optimize for net-zero goals.
2. **AI-driven material discovery and design with sustainability constraints:** The book acknowledges the role of AI and ML in process optimization. However, future research should heavily focus on developing more sophisticated AI and ML algorithms that can fundamentally accelerate the discovery and de novo design of novel materials. Crucially, these algorithms should be inherently designed to incorporate sustainability metrics as primary design constraints from the outset, rather than as post-design considerations. This means integrating factors like embodied carbon, toxicity, resource depletion potential, and recyclability into the AI's learning objectives for material synthesis. This also includes exploring advanced generative AI models capable of proposing entirely new material structures with predefined properties and environmental profiles, moving beyond mere optimization to true material innovation.
3. **Standardization and interoperability of digital platforms for net zero:** The diverse nature of industrial applications and the proliferation of various

digital tools, as highlighted across different chapters, necessitate extensive research into establishing common data standards, communication protocols, and truly interoperable digital platforms. This is critical to facilitate seamless and secure data exchange between disparate systems and across all stakeholders in the value chain—from material suppliers and manufacturers to logistics providers and end-users. Such standardization would enable a more holistic and integrated approach to achieving net-zero goals, breaking down current data silos, and fostering greater collaboration and transparency across the entire materials ecosystem.

4. **Socio-economic impact and policy frameworks for digitalized decarbonization:** While the book addresses technical and economic barriers, there is a pressing need for further comprehensive investigation into the broader socio-economic implications of widespread digitalization in the context of materials for net zero. This includes in-depth research on the necessary workforce training and reskilling initiatives required for the new digital landscape, the potential for new job creation, and ensuring equitable access to digital technologies, particularly for MSMEs, as discussed in Chapter 9. Furthermore, extensive research is required to develop robust and adaptive policy and regulatory frameworks that not only incentivize digital adoption but also actively support and accelerate investment in sustainable material innovations and their industrial implementation.
5. **Life cycle assessment (LCA) enhancement with digital tools:** The concept of environmental impact is central to net zero. Future research should focus on how real-time data generated from digitally connected manufacturing processes and supply chains can be integrated into dynamic and continuously updated LCA models. This would provide significantly more accurate and timely assessments of the environmental impacts of materials and products. Developing methodologies and automated tools to integrate LCA seamlessly into everyday material and product development workflows, offering continuous feedback for design and process improvements, is a key area for advancing sustainability.
6. **Scalability of Digital Solutions for Carbon Capture and Storage (CCUS):** Chapter 6 highlighted the significant challenges in scaling up NG-CCMs. Therefore, future research must intensely explore how advanced digital solutions can directly address the technical and economic barriers to deploying CCUS technologies at an industrial scale. This includes the application of sophisticated process control systems, predictive analytics for performance optimization and fault detection, and the development of high-fidelity digital twins for entire CCUS plants, enabling better decision-making, reduced operational costs, and accelerated deployment.
7. **Circular economy and digitalization of material flow:** The principles of a circular economy are fundamental to achieving net zero. Future research should focus on how digitalization can facilitate a more robust and efficient circular economy for materials used in net-zero applications. This encompasses developing advanced digital platforms for comprehensive material tracking throughout their life cycle, ensuring enhanced supply

chain transparency, and optimizing processes for material recycling, reuse, and remanufacturing. Ultimately, the goal is to leverage digital tools to minimize waste generation and maximize resource efficiency across all industrial sectors.

By pursuing these diverse yet interconnected research directions, the synergistic relationship between digitalization and material innovation can be profoundly strengthened, paving the way for a more resilient, sustainable, and truly net-zero future.

Index

O

P

R

S

V

For Product Safety Concerns and Information please contact our EU
representative GPSR@taylorandfrancis.com
Taylor & Francis Verlag GmbH, Kaufingerstraße 24, 80331 München, Germany

www.ingramcontent.com/pod-product-compliance
Ingram Content Group UK Ltd.
Pitfield, Milton Keynes, MK11 3LW, UK
UKHW020204200926
475702UK00004B/608

* 9 7 8 1 0 3 2 9 4 9 0 3 1 *